The Philosophy of S

The Philosophy of Software
Code and Mediation in the Digital Age

David M. Berry
University of Sussex, UK

© David M. Berry 2011, 2015

All rights reserved. No reproduction, copy or transmission of this publication may be made without written permission.

No portion of this publication may be reproduced, copied or transmitted save with written permission or in accordance with the provisions of the Copyright, Designs and Patents Act 1988, or under the terms of any licence permitting limited copying issued by the Copyright Licensing Agency, Saffron House, 6–10 Kirby Street, London EC1N 8TS.

Any person who does any unauthorized act in relation to this publication may be liable to criminal prosecution and civil claims for damages.

The author has asserted his right to be identified as the author of this work in accordance with the Copyright, Designs and Patents Act 1988.

First published 2011
Published in paperback 2015 by
PALGRAVE MACMILLAN

Palgrave Macmillan in the UK is an imprint of Macmillan Publishers Limited, registered in England, company number 785998, of Houndmills, Basingstoke, Hampshire RG21 6XS.

Palgrave Macmillan in the US is a division of St Martin's Press LLC, 175 Fifth Avenue, New York, NY 10010.

Palgrave is the global academic imprint of the above companies and has companies and representatives throughout the world.

Palgrave® and Macmillan® are registered trademarks in the United States, the United Kingdom, Europe and other countries.

ISBN 978–0–230–24418–4   hardback
ISBN 978–1–137–49027–8   paperback

A catalogue record for this book is available from the British Library.

A catalog record for this book is available from the Library of Congress.

Typeset by MPS Limited, Chennai, India.

Transferred to Digital Printing in 2015

For Trine

Contents

List of Figures	viii
Acknowledgements	x
1 The Idea of Code	**1**
Understanding computation	10
Towards digital humanities	18
2 What Is Code?	**29**
Code	33
Towards a grammar of code	51
Web 2.0 and network code	56
Understanding code	61
3 Reading and Writing Code	**64**
Tests of strength	65
Reading code	68
Writing code	75
Obfuscated code examples	86
4 Running Code	**94**
The temporality of code	97
The spatiality of code	98
Reverse remediation	99
Running code and the political	107
5 Towards a Phenomenology of Computation	**119**
Phenomenology and computation	127
The computational image	131
6 Real-Time Streams	**142**
Being a good stream	150
Financial streams	156
Lifestreams	162
Subterranean streams	167
Notes	172
Bibliography	182
Index	197

List of Figures

2.1	'Listen' by Sharon Hopkins	30
2.2	An example of 'beautiful' code as a sorting algorithm	48
2.3	'Rush' by Sharon Hopkins	49
2.4	The key differences between Web 1.0 and Web 2.0	57
3.1	Microsoft Windows source code commentary	69
3.2	Microsoft Windows source code 'moron' comments	69
3.3	Microsoft Windows source code 'hack' comments	70
3.4	Microsoft Windows source code 'undocumented' comments	71
3.5	Parody of the Microsoft Windows source code	72
3.6	Redacting command line execution	77
3.7	Underhanded C Contest, winning entry by John Meacham	78
3.8	Underhanded C Contest, contents are wiped keeping 255 as '000' length, showing how the basic image information is retained after redaction	79
3.9	Underhanded C Contest, second place entry by Avinash Baliga	80
3.10	Underhanded C Contest, third place entry by Linus Akesson	81
3.11	Simple example of a C program	83
3.12	C program with obfuscated characters with function call	84
3.13	C program now obfuscated through text changes and confusing formatting	85
3.14	Performs OCR of numbers 8, 9, 10 and 11	87
3.15	Prints spiralling numbers, laid out in columns	87
3.16	Maze displayer/navigator with only line-of-sight visibility	88
3.17	Computes arbitrary-precision square root	89
3.18	Makes X mouse pointer have inertia or anti-inertia	91
3.19	Flight simulator written in 1536 bytes of real code	92

4.1	Assembly language version of 'Hello, world!'	95
4.2	Binary file version of the executable	96
4.3	*Jaiken-zan*, each output is a combination of A and B	106
4.4	User represented in source code	115
4.5	'Voter' represented in the source code	115
4.6	The male 'voter' represented in the source code	116
4.7	The choice of the voter is technically constrained to only one candidate as represented in the source code	116

Acknowledgements

Writing remains to me an unusual practice that transforms my experience of the world whilst under the spell of writing. This book has had a particularly intensive birth, written as it is in the middle of the academic year and with everyday life swirling around it with all the attendant distractions. It has emerged from a number of related research themes that continue to guide my work and are focused on the challenge to thinking that is posed by technology. This work has been influenced, inspired, guided and challenged by such a plethora of authors that it is not possible to list them all here. However, I feel that they are all flowing in different modulations and intensities through the text that follows. I pass on this text in the hope that future readers will find something interesting in a subject I continue to find deeply fascinating.

I would like to take this opportunity to thank Nikki Cooper, the Callaghan Centre for the Study of Conflict, Power, and Empire, and the Research Institute for Arts and Humanities (RIAH) at Swansea University for funding the workshop, The Computational Turn, which explored key issues around software. Thanks also to N. Katherine Hayles and Lev Manovich and the other participants at this workshop who enthusiastically discussed many of the themes pertinent to this book in a rigorous and critical setting. I would also like to thank the many people who gave comments and suggestions to the text as it developed. In particular, Chapter 5 was presented at a number of places which assisted in writing, and so I would like to thank colleagues in the Department of Political and Cultural Studies at Swansea University and in particular Alan Finlayson and Roland Axtmann; the Department of Media and Film at Sussex University, particularly Michael Bull, Caroline Bassett, Sharif Mowlabocus, and Kate O'Riordan; The Law and Literature Association of Australia (LLAA) and The Law and Society Association of Australia and New Zealand (LSAANZ) and Griffith University for inviting me to present Chapter 5 in Brisbane, in particular William MacNeil; and lastly, Daniel Hourigan, Steve Fuller, Peter Bloom, William Merrin, and John Tucker for helpful additional comments. An early version of Chapter 6 was previously presented at Generation Net: Arts and Culture in the 21st century at Nottingham University, funded by the Institute of Film and Television Studies, and I would like to thank Iain Robert Smith for the invitation. A slightly reworked version of Chapter 4 was presented at Swansea

University in the Politics Research in Progress seminar series arranged by Jonathan Bradbury and I would like to thank all colleagues who attended for their generous feedback and ideas. Lastly, parts of Chapter 3 were presented at the New Materialisms and Digital Culture: An International Symposium on Contemporary Arts, Media and Cultural Theory at Anglia Ruskin University, and I would like to thank Jussi Parikka and Milla Tiainen for their invitation. I would also like to make a special note of thanks to Trine Bjørkmann Berry for reading and correcting early drafts of the chapters.

This book would not have been possible without the support and generosity of a great number of friends and colleagues at Swansea University who were always available to discuss subjects I found interesting. In particular, Claes Belfrage and Christian De Cock and the participants in the Cultural Political Economy research group, who may not realise that many of the ideas in the book were also aired there. I would also like to thank students on the MA Digital Media and my PhD students: Faustin Chongombe, Leighton Evans, Mostyn Jones, and Sian Rees for their useful contributions and discussions over the course of the year. Finally, I would like to thank my wife, Trine, and my children Helene, Henrik Isak, and Hedda Emilie, for waiting patiently, seemingly forever, to go to the beach.

DMB

1
The Idea of Code

Whilst we are dead to the world at night, networks of machines silently and repetitively exchange data. They monitor, control and assess the world using electronic sensors, updating lists and databases, calculating and recalculating their models to produce reports, predictions and warnings. In the swirling constellations of data, they oversee and stabilise the everyday lives of individuals, groups and organisations, and remain alert for criminal patterns, abnormal behaviour, and outliers in programmed statistical models. During our waking hours, a multitude of machines open and close gates and doors, move traffic-lights from red to green, and back to red again, monitor and authorise (or fail to authorise) our shopping on credit and debit cards, and generally keep the world moving. To do this requires millions, if not, billions of lines of computer code, many thousands of man-hours of work, and constant maintenance and technical support to keep it all running. These technical systems control and organise networks that increasingly permeate our society, whether financial, telecommunications, roads, food, energy, logistics, defence, water, legal or governmental. The amount of data that is now recorded and collated by these technical devices is astronomical. For example, 'Wal-Mart, a retail giant, handles more than 1 million customer transactions every hour, feeding databases estimated at more than 2.5 petabytes – the equivalent of 167 times the books in America's Library of Congress' and Facebook, a social-networking website, has collected 40 billion photos in its databases from the individual uploading of its users (*The Economist* 2010c). Search engines scour the web and deal with massive amounts of data to provide search results in seconds to users, with Google alone handling 35,000 search queries every second (*The Economist* 2010e). Significantly, '"information created by machines and used by other machines will probably grow faster than anything else", explains Roger Bohn of the UCSD, one of the authors of [a] study on American households. "This is primarily 'database to database' information—people are only tangentially involved in most of it"' (*The Economist* 2010d).

Of course, we have always relied upon the background activity of a number of bureaucratic processes for assigning, sorting, sending, and receiving information that have enabled modern society to function. But the specific difference introduced by software/code is that it not only increases the speed and volume of these processes, it also introduces some novel dimensions: (1) in a way that is completely new, software allows the delegation of mental processes of high sophistication into computational systems. This instils a greater degree of agency into the technical devices than could have been possible with mechanical systems;[1] (2) networked software, in particular, encourages a communicative environment of rapidly changing feedback mechanisms that tie humans and non-humans together into new aggregates. These then perform tasks, undertake incredible calculative feats, and mobilise and develop ideas at a much higher intensity than in a non-networked environment;[2] (3) there is a greater use of embedded and quasi-visible technologies, leading to a rapid growth in the amount of quantification that is taking place in society. Indeed, algorithms are increasingly quantifying and measuring our social and everyday lives. By capturing, in millions of different ways, the way we live, speak, act and think on mobile phones, CCTV cameras, websites, etc. computational devices are able to count these activities. This turns life into quantifiable metrics that are now visible and amenable for computation and processing. To give an idea of the extent to which computing power has grown, in 2010, using standard off-the-shelf hardware, 'computer scientists from the University of California, San Diego broke "the terabyte barrier" – and a world record – when they sorted more than one terabyte of data (1,000 gigabytes or 1 million megabytes) in just 60 seconds' (BJS 2010). This is roughly equivalent to the data on 40 single-layer Blu-Ray discs, 210 single-layer DVDs, 120 dual-layer DVDs or 1422 CDs (assuming CDs are 703 MB). Large collections of social aggregated data can easily exceed this size, so faster processing speeds are crucial for them to be data-mined for predictive, marketing, and social monitoring purposes by governments, corporations, and other large organisations, often without our knowledge or consent. This transforms our everyday lives into data, a resource to be used by others, usually for profit, which Heidegger terms *standing-reserve* (Heidegger 1993a: 322).

Computers are entangled with our lives in a multitude of different, contradictory and complex ways, providing us with a social milieu that allows us to live in a society that increasingly depends on information and knowledge. More accurately, we might describe it as a society that is more dependent on the computation of information, a *computational* knowledge society. Today, people rarely use the raw data, but consume it in processed form, relying on computers to aggregate or simplify the results for them, whether in financial credit-management systems, fly-by-wire aeroplanes, or expert-systems in medical diagnosis and analysis (*The Economist* 2010f). If we were to turn off the computers that manage these networks, the complexity of the modern world would come crashing in some cases, quite literally, to an abrupt halt. And yet, this is not the whole story, for each of the computers and technologies is actually mediating its own relationship with the world through the panoply of software. These computers run software that is spun like webs, invisibly around us, organising, controlling, monitoring and processing. As Weiner (1994: xv) says 'the growing use of software... represents a social experiment'.

Software is a tangle, a knot, which ties together the physical and the ephemeral, the material and the ethereal, into a multi-linear ensemble that can be controlled and directed. From the mundane activities of alarm clocks and heating systems, to complex structures like stock market trading systems and electricity grid markets, software helps these material objects function. But software can also change the very nature of what is considered possible: from the ability of terrestrial transmission networks to broadcast hundreds of simultaneous television channels and radio, as opposed to the previous small numbers of TV channels that analogue broadcasts enabled. Software can revolutionise the limitations of the physical world. In the case of care for Neonatal premature babies, for example, 'software ingests a constant stream of biomedical data, such as heart rate and respiration, along with environmental data gathered from advanced sensors and more traditional monitoring equipment on and around the babies' to compute real-time clinical updates on each child's physiological data streams to assist doctors in assessing their health' (IBM 2008). Software enables the fourth-generation jet fighters, like the Eurofighter Typhoon or the F16 Fighting Falcon, to be more effective fighter aircraft because they are deliberately designed to be aerodynamically unstable, a 'relaxed stability' design.[3] They can only be flown through the support of computers and software that manages their fly-by-wire systems; as 'F-16 pilots say, "You don't fly an F-16; it flies you", refer[ing] to the seemingly magical oversight of the electronic system' (Greenwood 2007).[4] Lastly, software underwrites the Internet itself, of course, which enables technical devices to communicate via special software-enabled protocols which construct an online world, the Web, from a complex constellation of different hardware, systems, telecommunications lines and devices.

But this software is too often hidden behind a façade of flashing lights, deceptively simple graphic user interfaces (GUIs) and sleekly designed electronic gadgets that re-presents a world to the user. As Kittler explains, '[s]ound and image, voice and text have become mere effects on the surface, or, to put it better, the interface for the consumer' (Kittler 1987: 102). But even at the level of the interface, software exceeds our ability to place limits on its entanglement, for it has in the past decade entered the everyday home through electronic augmentation that has replaced the mechanical world of the 20th century. From washing machines to central heating systems, to children's toys, television and video; the old electro-magnetic and servo-mechanical world is being revolutionised by the silent logic of virtual devices. It is time, therefore, to examine our virtual situation.[5]

As software increasingly structures the contemporary world, curiously, it also withdraws, and becomes harder and harder for us to focus on as it is embedded, hidden, off-shored or merely forgotten about. The challenge is to bring software back into visibility so that we can pay attention to both what it is (ontology), where it has come from (through media archaeology and genealogy) but also what it is doing (through a form of mechanology), so we can understand this 'dynamic of organized inorganic matter' (Stiegler 1998: 84).

Thankfully, software is also starting to become a focus of scholarly research from a variety of approaches loosely grouped around the field of software studies/cultural analytics (Fuller 2003; Manovich 2001, 2008) and critical code studies (Marino 2006; Montfort 2009). Some of the most interesting developments in this area include: *platform studies* (Montfort and Bogost 2009), where there is a critical engagement with an 'infrastructure that supports the design and use of particular applications, be it computer hardware, operating systems, gaming devices, mobile devices, and digital disc formats' (Gillespie 2008); *media archaeology*, which uncovers histories and genealogies of media, insisting on researching differences rather than continuity (Parikka 2007); research into *software engines*, which form an increasing part of the way in which software is packaged to perform a wide variety of functions, e.g. gaming engines, search engines, etc. (Helmond 2008); research into 'soft' authorship (Huber 2008) and genre analysis of software (Douglas 2008), which look at the way in which the notion of the author is problematised by the processual and bricolage nature of software development; graphical user interfaces, which focuses on the human–computer interface and the machine (Bratton 2008; Chun 2008; Dix et al 2003; Manovich 2001; Manovich and Douglas 2009); digital code literacy, which investigates how people read and write digital forms (Hayles 2004; Hayles 2005; Montfort 2008); research into temporality and code (Raley 2008); the sociology and political economy of the free software and open source movement, particularly with respect to the way in which the software has been rethought and subject to deliberation and contestation (Berry 2008; Chopra and Dexter 2008; Coleman 2009; Kelty 2008; Lessig 2002; May 2006; Weber 2005).[6]

Additionally, there has been important work in medium theory (Bassett 2007; Galloway 2006; Hayles 2007; Hansen 2006; Kittler 1997; Mackenzie 2006), critical attention to the creative industries (Gill and Pratt 2008; Garnham 2005: 26–7; Hesmondhalgh 2009; Kennedy 2010; Ross 2008), and attention to the theoretical challenge of digital media to media studies through web studies/media 2.0 and web science (Gauntlett 2009; Merrin 2009).[7] Within the field of political economy, too, there have been scholars looking at some of the important issues around software, although they have tended to focus on intellectual property rights (IPRs) (Benkler 2002, 2004, 2006; May 2006; Perelman 2002; Sell 2003), communications (Benkler 2006; McChesney 2007; Mosco 2009), or information (Benkler 2003a; Drahos and Braithwaite 2003; Mosco 1988) rather than the specific level of computer code itself.

All of these scholars are in some way exploring the phenomena of computer code from a number of disciplinary perspectives, even if indirectly. What remains clear, however, is that looking at computer code is difficult due to its ephemeral nature, the high technical skills required of the researcher and the lack of analytical or methodological tools available. This book will attempt to address this lack in the field by pointing towards different ways of understanding code. It will do so through a phenomenological approach that tries to highlight the *pragmata* of code. Following Fuller (2008), it will attempt to:

> Show the stuff of software in some of the many ways that it exists, in which it is experienced and thought through, and to show, by the interplay of concrete examples and multiple kinds of accounts, the condition of possibility that software establishes (Fuller 2008: 1).

The book is also intended to be a critical introduction to the complex field of understanding digital culture and technology, offering a way into the subject for those in the humanities/social sciences or the digital humanities. Indeed, I argue that to understand the contemporary world, and the everyday practices that populate it, we need a corresponding focus on the computer code that is entangled with all aspects of our lives, including reflexivity about how much code is infiltrating the academy itself. As Fuller (2006) argues, 'in a sense, all intellectual work is now "software study", in that software provides its media and its context... [yet] there are very few places where the specific nature, the materiality, of software is studied except as a matter of engineering'. We also need to think carefully about the 'structure of feeling' that computer code facilitates and the way in which people use software in their everyday lives and practices. For example, this includes the increase in people's acceptance and use of: life-style software, e.g. Nike+; personal mobility software, e.g. GPS and SatNav; cultural software, e.g. photoshop and InDesign (Manovich 2008); gaming, both console and networked (Wark 2007); 'geo' or location, e.g. Gowalla and FourSquare; and social media, such as Facebook, Twitter, QQ, TaoTao, etc.

The way in which these technologies are recording data about individuals and groups is remarkable, both in terms of the collection of: (1) formal technical data, such as dataflows, times and dates, IP addresses, geographical information, prices, purchases and preferences, etc.; (2) but also qualitative feelings and experiences. These *software avidities* are demonstrated when Twitter asks the user: 'What's happening?', Facebook asks: 'What's on your mind?', and Google Buzz inquires: 'Share what you're thinking'. This information is not collected passively, but processed and related back into the experience of other users either as a news feed or data stream, or else as an information support for further use of the software. Amazon uses this raw data about what other readers have bought to provide information back to users as personalised recommendations to drive new sales. Google, on the other hand, generates personalised adverts and pre-populates its search boxes with a function called 'Google Query Suggestions'.[8] When one types 'What happens when' into the Google search-box, you are presented with a pre-computed list of the most popular questions typed by users into the search engine, and as of June 2010 they were:

What happens when you die
What happens when you lose your viginity
What happens when you stop smoking

What happens when you have a miscarriage
What happens when a volcano erupts
What happens when you have an abortion
What happens when we die
What happens when you deactivate facebook account
What happens when there is a hung parliament
What happens when a country goes bankrupt

This is the result of a massive computational analysis of people's search texts and page rankings using statistical algorithms. Indeed, these results certainly reflect a number of issues of the time, such as the eruption of Iceland's *Eyjafjallajökull* volcano, public disquiet with Facebook privacy, the last election in the UK which delivered a hung parliament and a Conservative and Liberal coalition, and the financial crisis of 2007–10.[9] However, without an understanding of how computation is tying data, news, practices and search results together through computer code, the process of 'search' is difficult to explain, if not strangely magical. It also precludes us from concentrating on the political economic issues raised by the fact that an American corporation is capturing this data in the first place, and is able to feed it back through pre-populating the search box and hence steer people in particular directions. Google has a reported 98 per cent of the mobile search market and 71 per cent of the search share market globally (Pingdom 2010).[10] Indeed '[i]n the process of ranking results, search engines effectively create winners and losers on the web as a whole' (Halavais 2008:85). For example, the ability to pay to be on the Google pre-populated search list would presumably be a desirable way of advertising and driving sales. This is the continuing logic of Google's business model that is an exemplar of Smythe's (2006) notion of the 'audience commodity'. Essentially, Google creates advertising markets by the real-time segmentation of search requiring computer power to understand who, in terms of a particular consumer group, is searching and what can be advertised to them. Google, therefore, harnesses the power of computation to drive an aggressive advertising strategy to find out who the user is, and what are their preferences, tastes, desires, and wants. Indeed, '[r]ivals have accused Google of placing the Web sites of affiliates... at the top of Internet searches and relegating competitors to obscurity down the list' (*New York Times* 2010). Indeed, Google was recently awarded a patent on displaying search results based on how the users moves their mouse cursor on the screen allowing it to monitor user behaviour at a very low level of granularity, the so-called click-stream. This raises serious privacy concerns as the collection of such statistics could be used for analysis of users 'pre-cognition' and then tailored for behavioural marketing (Wilson 2010). It is clear that access to this kind of data and analysis, as a service from Google, could be extremely valuable to advertisers. Indeed, Eric Schmidt, CEO of Google, recently commented:

> "I actually think most people don't want Google to answer their questions," he elaborates. "They want Google to tell them what they should be doing next." Let's say you're walking down the street. Because of the info Google has collected about you, "we know roughly who you are, roughly what you care about, roughly who your friends are." Google also knows, to within a foot, where you are. Mr. Schmidt leaves it to a listener to imagine the possibilities: If you need milk and there's a place nearby to get milk, Google will remind you to get milk. It will tell you a store ahead has a collection of horse-racing posters, that a 19th-century murder you've been reading about took place on the next block. Says Mr. Schmidt, a generation of powerful handheld devices is just around the corner that will be adept at surprising you with information that you didn't know you wanted to know. "The thing that makes newspapers so fundamentally fascinating – that serendipity – can be calculated now. We can actually produce it electronically," Mr. Schmidt says (Jenkins 2010).

Whilst it might be tempting to think that the question of code, or perhaps better the *politics of code*, is of little importance, one should remember that Google had revenues of $17 billion in 2007, $22 billion in 2008, and $24 billion in 2009 (Google 2010b). They also represent an important unregulated gateway into the information contained upon the web, and governments, amongst others, have been concerned with their growing informational and economic power. For example, in *Viacom vs Google 2010*, when Viacom sued Google for copyright infringement over the use of its content on Youtube, Google spent over $100 million on lawyers to defend the case, and was ultimately successful. Google now has a remarkable $30.1 billion in cash, cash equivalents, and short-term securities that it can use to defend itself and spend on developing new products and services (Sieglar 2010). Google has responded to these critics who argue that there should be 'search neutrality' by saying,

> The world of search has developed a system in which each search engine uses different algorithms, and with many search engines to

choose from users elect to use the engine whose algorithm approximates to their personal notion of what is best for the task at hand. The proponents of "search neutrality" want to put an end to this system, introducing a new set of rules in which governments would regulate search results to ensure they are fair or neutral (Mayer 2010).

To understand these kinds of issues, which are essentially about the regulation of computer code itself, we need to be able to unpack the way in which these systems are built and run. This means a closer attention to the multiple ways in which code is deployed and used in society. This can take place on a number of levels, for the social researcher: through reading code there may be a method for uncovering patterns in current worries, questions, issues and debates taking place in a society at any one time, perhaps a form of technological unconscious shown in search results. For the political economist it can help demonstrate the way in which economic power is now manifested, perhaps through the control and use of code libraries, patents and digital rights management technologies. Understanding code can also reveals ethical questions, both in terms of our relationship to ourselves as autonomous beings, but also the framing and homogenisation of ideas and practice – heteronomy versus autonomy.[11] A close reading of code can also draw attention to the way in which code may encode particular values and norms (see Berry 2008: 31) or drive particular political policies or practices.

Therefore, it seems to me that we need to become more adept at *reading* and *writing* computer code in order to fully understand the technical experience of the everyday world today. This knowledge would also allow us to decode the more rarefied worlds of high technology, finance, politics and international political economy, to name just a few examples. Without this expertise, when tracing the agentic path, whether from cause to effect, or through the narrative arcs that are used to explain our contemporary lives, we will miss a crucial translation involved in the technical mediation provided by software. As Mackenzie (2003) perceptively puts it:

> code runs deep in the increasingly informatically regulated infrastructures and built environments we move through and inhabit. Code participates heavily in contemporary economic, cultural, political, military and govermental domains. It modulates relations within collective life. It organises and disrupts relations of power. It alters the conditions of perception, commodification and representation (Mackenzie 2003: 3).

Computer code needs to be analysed not only as a technology, but also as a medium materialised into particular code-based devices. Therefore I want to argue that a powerful way to reconceptualise code is through the notion of a *super-medium* (Berry 2008: 34, *cf* Manovich 2008: 79–80), that is, that code unifies the fragmented mediums of the twentieth century (tv/film/radio/print) within the structures of code (using Kittler's notion of the implosion of media forms). Code is not a medium that *contains* the other mediums, rather it is a medium that radically reshapes and transforms them into a new unitary form. This super-medium acts as both a mediating and structurating frame that we must understand through its instantiation under particular physical constraints.

That is, I reject the so-called 'immateriality' of software and draw attention to the concrete thing-in-the-world-ness of software so that we can 'see what it is, what it does, and what it can be coupled with' (Fuller, M. 2008: 3). However, it is also clear that we have not yet found adequate means to analyse the multifaceted dimensions to code. To understand code and place the question of code firmly within the larger political and social issues that are raised in contemporary society, we need to pay more attention to the *computational ontology* of code. In other words, the way in which code is actually 'doing' is vitally important, and we can only understand this by actually reading the code itself and watching how it operates. As a contribution to developing our understanding of what is admittedly a complex subject, I take a synoptic look at the phenomena of code, and try to place it within phenomenological context to understand the profound ways in which computational devices are changing the way in which we run our politics, societies, economies, the media and even everyday life. Throughout the book, then, I will explore code by looking at the assemblage presented by the computational and the human. In particular, the way in which our relationships with the many entities that populate this human-built world are increasingly embedded with digital microprocessors running digital code. This universality of code within all manner of devices and many different fields of knowledge raises important questions for all disciplines and research fields, something that will be a running theme throughout the book.

Understanding computation

The term computation itself comes from the Latin *computare*, *com*- 'together' and *putare* 'to reckon, to think or to section to compare the pieces'. To compute, then, is to 'to count, or to calculate'. For computer scientists, computation (or information processing) is a field of research that investigates what can and what cannot be calculated. Closely allied with this is a certain comportment towards the world maintained through computational skills and techniques. My intention is not to evaluate or outline the theoretical underpinnings of computability as a field within the discipline of computer science, rather, I want to understand how our being-in-the-world, the way in which we act towards the world, is made possible through the application of these theoretical computational techniques, which are manifested in the processes, structures and ideas stabilised by software and code.

It is also worth clarifying that I do not refer to computational in terms of *computationalism*, a relatively recent doctrine that emerged in analytic philosophy in the mid 1990s, and which argues that the human mind is ultimately 'characteristable as a kind of computer' (Golumbia 2009: 8), or that an increasing portion of the human and social world is explainable through computational processes.[12] This is what Hayles (2005) calls the Regime of Computation, whereby

> [it] provides a narrative that accounts for the evolution of the universe, life, mind, and mind reflecting on mind by connecting these emergences with computational processes that operate both in human-created simulations and in the universe understood as software running on the "Universal Computer" we call reality (Hayles 2005: 27).

Additionally, some theorists of computation quickly move from their theoretical and empirical work to the speculative, hence they claim that the universe is digital all the way down. One example is Edward Fredkin, who proposes a form of 'digital philosophy', that argues 'that the discrete nature of elementary particles indicates that the universe is discrete, rather than continuous, digital rather than analog' (Hayles 2005: 23).[13] This is the idea that all we need to do is uncover the digital rules that underlie reality (Borgmann 1999: 11). As Hayles observes,

> The regime [of computation] reduces ontological requirements to a bare minimum. Rather than an initial premise (such as God, an originary Logos, or the axioms of Euclidean geometry) out of which multiple entailments spin, computation requires only an elementary distinction between something and nothing (one and zero) and a small set of logical operations... far from presuming the "transcendental signified" that Derrida identifies as intrinsic to classic

metaphysics, computation privileges the emergence of complexity from simple elements and rules (Hayles 2005: 23).

More speculatively, computability, within the discipline of computer science, is now seen to also include *non-discrete* continuous data, called analogue computing platforms.[14] The common example given is that of a spaghetti sorter, which is actualised through the action of banging spaghetti of different lengths on the flat surface of the table. This causes multiple 'processing' of the material elements (i.e. spaghetti strands) which results in the spaghetti being sorted by size (that is, sorts a list of n numbers in order n time) (Chalmers 1989). Here, there is an input, a processing dimension, and an output, which has not been sent on a detour through a digital device. This is interesting due to the way in which the world outside is cast, as shown by Beggs et al. (2009), who refer to this computational relationship with the external world (which is problematised as being potentially non-computational) as physical experimentation with *oracles*. Here, physical experiments in the universe are understood as 'oracles' to algorithms, that is, the external world is a problematised space for the computational which requires an interface through which the oracle (i.e. the physical world) can be consulted by the algorithm. True to their Delphic forebears, oracles are positioned as risky and may yield results that are essentially non-computational and which may need to be tamed through the use of protocols and interfaces between the computational and non-computational world – in a sense the placing of a computational filter on the world.[15]

Although I will not be looking in detail at the questions raised by analogue computation, nor the digital philosophy of Fredkin and others, these examples demonstrate the increasing importance of the digital in how people are conceptualising the world. Certainly, the growth in importance of a computational comportment is connected to the unparalleled rise in importance of computers and technology. One is drawn to the analogy,

> with eighteenth-century commentators who, impressed by the reductive power of Newton's laws of motion and the increasing sophistication of time-keeping mechanisms, proclaimed that the universe was a clockwork (Hayles 2005: 3).

These metaphors help us understand the world, and with a shift to computational metaphors, certain aspects of reality come to the fore, such as the notion of orderliness, calculability, and predictability, whilst others, like chaos, desire and uncertainty, retreat into obscurity. One might speculate, for example, of the extent to which constructions of subjectivity expressed in the humanities retreat when 'one' is no longer 'hallucinating a meaning between letters and lines' when reading books (Kittler, quoted in Hayles 2005: 4) and is instead a part of a network of instant messages, emails and datascapes in a multi-visual media ecology. This is certainly indicated in the importance, for Heidegger, of questioning technology and understanding technology as an ontological condition for our comportment towards the world (Heidegger 1993a).

Many other attempts to understand the computational tend towards equating it with instrumental rationality (e.g. Golumbia 2009). In contrast, I argue for a distinction between computationalist and instrumentalist notions of reason. I use the definition of instrumental rationality as the application by an actor of *means* to *ends* through mathematics, empirical knowledge and logic. This is a notion of the maximisation of instrumentality in order to produce the maximum output for a given input, the classic example being the utility-maximising rationality of the individual selfish actor. In effect, instrumental rationality is a mode of reasoning employed by an agent. In contrast, computational rationality is a special sort of knowing, it is essentially vicarious, taking place within other actors or combinations and networks of actors (which may be human or non-human) and formally algorithmic. One thinks here of writing a poem within a word-processor, which appears to the computer as a constantly deferred process of manipulating symbolic data, and which the computer is never in a position to judge as a completed task – even when stored on a hard-disk it remains merely temporarily frozen between user edits. This means that the location of reasoning is highly distributed across technical devices and the agents. This strongly entangles the computational with the everyday world; after all, only a limited number of computational tasks are self-contained and have no user or world input.[16] This also points to the fact that computational rationality can be made up of different forms of rationality itself, not necessarily purely instrumental, including, for example communicative moments, aesthetic moments and expressive moments. In this sense then, computational rationality is a form of reasoning that takes place through other non-human objects, but these objects are themselves able to exert agential features, either by making calculations and decisions themselves, or by providing communicative support for the user.

Computational devices therefore have a potentially communicative dimension, as each technical actor must be in constant communication with the other actors for the computation to function (whether function, object, code, human, non-human, etc.). The computational device, as an algorithmic totality, is in a constant state of exception from multiple events which must be attended to, that is, the device is constantly interrupted by a parliament of things or users. The seemingly end-directed nature of computational processes may actually be constantly deferred internally, that is, never reaching a final goal. In a certain sense, this is an agonistic form of communicative action where devices are in a constant stream of data flow and decision-making which may only occasionally feedback to the human user.

This 'everyday computational' is a comportment towards the world that takes as its subject-matter everyday objects which it can transform through calculation and processing interventions. The definition of possible states and events is usually formulated in a computation model, such as the Turing machine or the finite state automata, and embedded in computational devices. For Stiegler (2009) this way of thinking about the world is epitomised in,

> the beginning of a *systematic discretization*... – that is to say, of a vast process of the *grammaticalization of the visible*. Just as, today, the language industries are producing digital dictionaries (which is to say, grammars), there are presently being realised [many new digital] "grammers" and "dictionaries"... These involve, in effect, simulations in physics, chemistry and astrophysics, simulations in training and ergonomics, virtual worlds, clones of real beings, artificial intelligence, form recognition, artificial life, and artificial death (Stiegler 2007: 149).

For computer scientists, it is the translation of the continuum into the discrete that marks the condition of possibility for computational ontologies. Only when things are turned into a digital form are they available to be manipulated through digital technology and computer code. Indeed, 'digital computation is fundamentally computation by algorithms, which operate on symbols in discrete time' (Tucker and Zucker 2007: 2).[17] This reminds us that computation is limited to specific temporal durations and symbolic sets of discrete data to represent reality, but once encoded, it can be resampled, transformed, and filtered endlessly.

This demonstrates the plasticity of digital forms and points toward a new way of working with representation and mediation, facilitating the digital folding of reality. To mediate an object, a computational device requires that it be translated. This minimal transformation is effected through the input mechanism of a socio-technical device within which a model or image is stabilised and attended to, and then internally transformed depending on a number of interventions, processes or filters and then outputted as a final calculation. This results in real-world situations where computation is event-driven and divided into discrete processes to undertake a particular user task. The key point is that without the possibility of *discrete* encoding there is no object for the computational device to process; however, in cutting up the world in this manner, information about the world necessarily has to be discarded in order to store a representation within the computer. In other words, a computer requires that everything is transformed from the continuous flow of our everyday reality into a grid of numbers that can be stored as a representation of reality which can then be manipulated using algorithms. The other side of the coin, of course, is that these subtractive methods of understanding reality (*episteme*) produce new knowledges and methods for the control of reality (*techne*).

For objects in the world to be computational requires that they offer a certain set of affordances facilitated through the operation of computer code. This is managed through the writing of code that determines certain functions that the software is engineered to perform. These can be at the level of the software itself, and hence invisible to the user directly (for example application programming interfaces or APIs), or presented to the user through a visual interface which allows the user to determine what it does, its *affordance*. To distinguish between the two, it is useful to think of *hidden* affordances and *visible* affordances. That is, with visible affordances,

> The value is clear *on the face of it*... The postbox "invites" the mailing of a letter, the handle "wants to be grasped", and things "tell us what to do with them" (Gibson 1977: 136).

In a similar way to physical objects, technical devices present to the user a certain function, or range of functions, that are stabilised and formatted through a particular human-computer interface, very often graphical. That is not to say that non-screenic affordances aren't used, they clearly are where the interface requires only a simple input from the user – think of the famous iPod wheel – but this set of functions (affordances) in a computational device is always a partial offering that may be withheld or remain unperformed. This is because the device has an internal state which is generally withheld from view and is often referred to as a 'black box', indicating that it is opaque to the outside viewer.[18]

In other words, the user has no way of knowing directly that their actions has had the result they desired, except as reported on the surface by the technical device.

In this sense then, the computational device is a mediator between entities and their phenomenal representation in the everyday world, and its affordances help inform us and guide us in using it. To manipulate the invisible or imperceptible informational entities we increasingly deal with in today's world, such as data, electronic money or objects at a distance, requires some form of computational mediation. For example, if a computer microscope displays microbes which are beyond the range of human sight and which therefore require translation through techniques that can magnify them, the magnification is undertaken through the computational manipulation of the input data using algorithms, rather than through a purely mechanical or optical process. What I am pointing towards here is the displaying of a representation, which could be manipulated in a number of different ways by the processing software before being displayed back to the user. Here, certain functions will be made available in the software that guide the user in particular ways (its affordances, such as increase magnification, decrease magnification, rotate slide, etc.), but due to the loose coupling of code and interface there is no guarantee that the representation on screen is actually undertaking the task we have commanded. We have to trust the machine has properly captured, transformed, and rendered the desired image.

If we consider the digital representation of a microbe, for example, there is a translation from a physical analogue microbe via a sensitive detector called an analogue-to-digital convertor, which provides a conversion to a digital form. This is then stored within the computer memory as a series of digital data points, a stream of numbers. These in turn can be processed and manipulated in a variety of ways by the computer, for example magnified, colour corrected, or analysed computationally to look for patterns. This new processed representation as a stream of numbers is then finally translated back onto the computer screen for the user and rendered as a screenic image. Of course, there is also the possibility of further interaction from the user to manipulate the data. However, at every stage of the process the user is reliant on the software to mediate this mediation as there is no other access to the data nor the transformations. This demonstrates the *double mediation* which makes the user increasingly reliant on the screen image that the computer produces, but also renders them powerless to prevent the introduction of errors and mistakes (unless they have access to the computer code).

Naturally, at any moment the computer may also introduce errors as part of its processing, in addition to these digital artefacts, screen image effects may be produced due to the limitations of the particular resolution chosen for the original conversion. This is a classic problem in health sciences, for example, where the doctor must quickly determine whether the shadow on a patient scan represents a medical issue or merely a computational artefact introduced by the process.

That is not to say matter too is not also the subject of feverish research activity into making it 'computable'. For example, researchers at MIT and Harvard have created a 'piece of paper that folds itself' into origami folds using actuators embedded in the material (Geere 2010b), and engineers exploring 'relationship between the architecture of spaces music is composed in and performed' have created what they term a Tunable Sound Cloud, which allows the space in a room to be dynamically modified either in response to sound, or to maximise the listening experience to sound and music (Geere 2010a). However, much of the code that we experience in our daily lives is presented through a visual interface that tends to be graphical and geometric, and where haptic, through touch, currently responds through rather static physical interfaces but dynamic visual ones, for example iPads, touch screen phones, etc.

Computer code is not solely technical though, and must be understood with respect to both the 'cultures of software' that produce it, but also the cultures of consumption that surround it. Users avidly purchase and use both its direct software products or tangentially through the production of goods and services that cultural software enables, for example games, websites, music, etc. Not forgetting, of course, the cultural insecurities that the computational processes instil in people more generally, especially when mediated through popular culture, for example, in music, in Lil B's track *The Age Of Information* (LilB 2010) and on film in *The Matrix* (The Wachowski Brothers, 1999). Therefore, following Kittler's (1997) definition of media, I also want to understand computational reasoning as a cultural technique, one that allows one to select, store, process, and produce data and signals for the purposes of various forms of action but with a concentration on its technical materiality (Kramer 2006: 93).

Thus, there is an undeniable cultural dimension to computation and the medial affordances of software. This connection again points to the importance of engaging with and understanding code, indeed, computer code can serve as an index of digital culture (imagine mapping different programming languages to the cultural possibilities that it affords, e.g. HTML to cyberculture, AJAX to social media).[19] This means that we can ask the question: what is culture after it has been 'softwarized'? (Manovich 2008:41). Understanding code can therefore be a resourceful way of understanding cultural production more generally, for example, digital typesetting transformed the print newspaper industry, eBook and eInk technologies are likely to do so again.

Towards digital humanities

By problematising computational ontologies, we are able to think critically about how knowledge in the 21st century is transformed into information through computational techniques, particularly within software. It is interesting that at a time when the idea of the university is itself under serious rethinking and renegotiation, digital technologies are transforming our ability to use and understand information outside of these traditional knowledge structures. This is connected to wider challenges to the traditional narratives that served as unifying ideas for the university and with their decline has led to difficulty in justifying and legitimating the post-modern university *vis-à-vis* government funding.

Historically, the role of the university has been closely associated with the production of knowledge. For example, Immanuel Kant outlined an argument for the nature of the university in 1798, called *The Conflict of the Faculties*. He argued that all of the university's activities should be organised by a single regulatory idea, that of the concept of reason. As Bill Readings (1996) argued:

> Reason on the one hand, provide[d] the *ratio* for all the disciplines; it [was] their organizing principle. On the other hand, reason [had] its own faculty, which Kant names[d] 'philosophy' but which we would now be more likely to call the 'humanities' (Readings 1996: 15).

Kant argued that reason and the state, knowledge and power, could be unified in the university by the production of individuals capable of rational thought and republican politics – the students trained for the civil service and society. Kant was concerned with the question of regulative public reason, that is, how to ensure stable, governed and governable regimes which can rule free people, in contrast to tradition represented by monarchy, the Church or a Leviathan. This required universities, as regulated knowledge-producing organisations, to be guided and overseen by the faculty of philosophy, which could ensure that the university remained rational. This was part of a response to the rise of print culture, growing literacy and the kinds of destabilising effects that this brought. Thus, without resorting to dogmatic doctrinal force or violence one could have a form of perpetual peace by the application of one's reason.[20]

This was followed by the development of the modern university in the 19th century, instituted by the German Idealists, such as Schiller and Humboldt, who argued that there should be a more explicitly political role to the structure given by Kant. They argued for the replacement of reason with culture, as they believed that culture could serve as a 'unifying function for the university' (Readings 1996: 15). For the German Idealists, like Humboldt, culture was the sum of all knowledge that is studied, as well as the cultivation and development of one's character as a result of that study. Indeed, Humboldt proposed the founding of a new university, the University of Berlin, as a mediator between national culture and the nation-state. Under the project of 'culture', the university would be required to undertake both research and teaching, respectively the production and dissemination of knowledge. The modern idea of a university, therefore, allowed it to become the preeminent institution that unified ethnic tradition and statist rationality by the production of an educated cultured individual. The German Idealists proposed,

> that the way to reintegrate the multiplicity of known facts into a unified cultural science is through *Bildung*, the enoblement of character... The university produces not servants but *subjects*. That is the point of the pedagogy of *Bildung*, which teaches knowledge acquisition as a *process* rather than the acquisition of knowledge as a product. (Readings 1996: 65–7).

This notion was given a particularly literary turn by the British, in particular John Henry Newman and Mathew Arnold, who argued that literature, not culture or philosophy, should be the central discipline in the university, and also of national culture more generally.[21] Literature, therefore, became institutionalised within the university 'in explicitly national terms and an organic vision of the possibility of a unified national culture' (Readings 1996: 16). This became regulated through the notion of a literary canon, which was taught to students to produce literary subjects as national subjects.

Readings (1996) argues that in the post-modern university we now see the breakdown of these ideals, associated particularly with the rise in the notion of the 'university of excellence' which he argues is a concept of the university that has no content, no referent. What I would like to suggest is that instead we are beginning to see the cultural importance of the digital as the unifying idea of the university. Initially this has tended to be associated with notions such as *information literacy* and *digital literacy*, betraying their debt to the previous literary conception of the university, albeit understood through vocational training and employment. However, I want to suggest that rather than learning a *practice* for the digital, which tends to be conceptualised in terms of ICT skills and competences (see for example the European Computer Driving License[22]), we should be thinking about what reading and writing actually should mean in a computational age. This is to argue for critical understanding of the *literature* of the digital, and through that develop a shared digital culture through a form of digital *Bildung*. Here I am not calling for a return to the humanities of the past, to use a phrase of Fuller (2010), 'for some humans', but rather to a liberal arts that is 'for all humans'. To use the distinction introduced by Hofstadter (1963), this is to call for the development of a digital *intellect* as opposed to a digital *intelligence*. He writes:

> Intellect... is the critical, creative, and contemplative side of mind. Whereas intelligence seeks to grasp, manipulate, re-order, adjust, intellect examines, ponders, wonders, theorizes, criticizes, imagines. Intelligence will seize the immediate meaning in a situation and evaluate it. Intellect evaluates evaluations, and looks for the meanings of situations as a whole... Intellect [is] a unique manifestation of human dignity (Hofstadter 1963: 25).

The digital assemblages that are now being built, not only promise great change at the level of the individual human actor. They provide destabilising amounts of knowledge and information that lack the regulating force of philosophy that, Kant argued, ensures that institutions remain rational. Technology enables access to the databanks of human knowledge from anywhere, disregarding and bypassing the traditional gatekeepers of knowledge in the state, the universities, and market. There no longer seems to be the professor who tells you what you should be looking up and the 'three arguments in favour of it' and the 'three arguments against it'. This introduces not only a moment of societal disorientation with individuals and institutions flooded with information, but also offer a computational solution to them in the form of computational rationalities, what Turing (1950) described as super-critical modes of thought. Both of these forces are underpinned at a deep structural level by the conditional of possibility suggested by computer code.

As mentioned previously, computer code enables new communicative processes, and with the increasing social dimension of networked media the possibility of new and exciting forms of collaborative thinking arises. This is not the collective intelligence discussed by Levy (1999), rather, it is the promise of a collective *intellect*. This is reminiscent of the medieval notion of the *universitatis*, but recast in a digital form, as a society or association of actors who can think critically together mediated through technology. It further raises the question of what new modes of collective knowledge software can enable or constitute. Can software and code take us beyond the individualising trends of blogs, comments, twitter feeds, and so forth, and make possible something truly collaborative? Something like the super-critical thinking that is generative of ideas, modes of thought, theories and new practices?

For the research and teaching disciplines within the university, the digital shift could represent the beginnings of a moment of 'revolutionary science', in the Kuhnian sense, of a shift in the ontology of the positive sciences and the emergence of a constellation of new 'normal science' (Kuhn 1996). This would mean that the disciplines's would, ontologically, have a very similar Lakatosian computational 'hard core' (Lakatos 1980).[23] This has much wider consequences for the notion of the unification of knowledge and the idea of the university (Readings 1996). Computer Science could play a foundational role with respect to the other sciences, supporting and directing their development, even issuing 'lucid directives for their inquiry' (see Thomson (2003: 531) for a discussion of how Heidegger understood this to be the role of philosophy). Perhaps we are beginning to see reading and writing computer code as part of the pedagogy required to create a new subject produced by the university, a *computational* or *data-centric* subject. This is, of course, not to advocate that the *existing* methods and practices of computer science become hegemonic, rather that a *humanistic* understanding of technology could be developed, which also involves an urgent inquiry into what is human about the *computational* humanities or social sciences. In a related manner, Steve Fuller (2006) has called for a 'new sociological imagination', pointing to the historical project of the social sciences that have been committed to 'all and only humans' because they 'take all human beings to be of equal epistemic interest and moral concern' (Fuller 2010: 242). By drawing attention to 'humanity's ontological precariousness' (ibid.: 244), Fuller rightly identifies that the project of humanity requires urgent thought, and we might add even more so in relation to the challenge of a *computational ontology* that threatens our understanding of what is required to be identified as human at all.

If software and code become the condition of possibility for unifying the multiple knowledges now produced in the university, then the ability to think oneself, taught by rote learning of methods, calculation, equations, readings, canons, processes, etc, might become less important. Although there might be less need for an *individual* ability to perform these mental feats or, perhaps, even recall the entire canon ourselves due to its size and scope, using technical devices, in conjunction with collaborative methods of working and studying, would enable a cognitively supported method instead. The internalisation of particular practices that have been instilled for hundreds of years would need to be rethought, and in doing so the commonality of thinking *qua* thinking produced by this pedagogy would also change. Instead, reasoning could change to more conceptual or communicational method of reasoning, for example, by bringing together comparative and communicative analysis from different disciplinary perspectives and knowing how to use technology to achieve a result that can be used – a rolling process of reflexive thinking and collaborative rethinking. Relying on technology in a more radically decentred way, depending on technical devices to fill in the blanks in our minds and to connect knowledge in new ways, would change our understanding of knowledge, wisdom and intelligence itself. It would be a radical decentring in some ways, as the Humboldtian subject filled with culture and a certain notion of rationality, would no longer exist, rather, the computational subject would know where to recall culture as and when it was needed in conjunction with computationally available others, a *just-in-time* cultural subject, perhaps, to feed into a certain form of connected *computationally* supported thinking through and visualised presentation. Rather than a method of thinking with eyes and hand, we would have a method of thinking with eyes and screen.[24] This stream-like subjectivity is discussed in detail later.

This doesn't have to be dehumanising. Latour and others have rightly identified the domestication of the human mind that took place with pen and paper (Latour 1986). This is because computers, like pen and paper, help to stabilise meaning, by cascading and visualising encoded knowledge that allows it to be continually 'drawn, written, [and] recoded' (Latour 1986: 16). Computational techniques could give us greater powers of thinking, larger reach for our imaginations, and, possibly, allow us to reconnect to political notions of equality and redistribution based on the potential of computation to give to each according to their need and to each according to their ability. This is the point made forcefully by Fuller (2010: 262) who argues that we should look critically at the potential for inequality created when new technologies are introduced into society. This is not merely a problem of a 'digital divide', but a more fundamental one of how we classify those that are more 'human' than others, when access to computation and information increasingly has to pass through the market.

The importance of understanding computational approaches is increasingly reflected across a number of disciplines, including the arts, humanities and social sciences, which use technologies to shift the critical ground of their concepts and theories – essentially a *computational turn*.[25] This is shown in the increasing interest in the *digital humanities* (Schreibman *et al.* 2008) and *computational social science* (Lazer *et al.* 2009), for example, the growth in journals, conferences, books and research funding. In the digital humanities 'critical inquiry involves the application of algorithmically facilitated search, retrieval, and critical process that, originating in humanities-based work', therefore 'exemplary tasks traditionally associated with humanities computing hold the digital representation of archival materials on a par with analysis or critical inquiry, as well as theories of analysis or critical inquiry originating in the study of those materials' (Schreibman *et al.* 2008: xxv). In computational social sciences, Lazer *et al.* (2009) argue that 'computational social science is emerging that leverages the capacity to collect and analyze data with an unprecedented breadth and depth and scale'.

Latour speculates that there is a trend in these informational cascades, which is certainly reflected in the ongoing digitalisation of arts, humanities and social science projects that tends towards 'the direction of the greater merging of figures, numbers and letters, merging greatly facilitated by their homogenous treatment as binary units in and by computers' (Latour 1986: 16). The financial considerations are also new with these computational disciplines, as they require more money and organisation than the old individual scholar of lore. Not only are the start-up costs correspondingly greater, usually to pay for the researchers, computer programmers, computer technology, software, digitisation costs, etc. but there are real questions about sustainability of digital projects, such as: who will pay to maintain the digital resources?, 'Will the user forums, and user contributions, continue to be monitored and moderated if we can't afford a staff member to do so? Will the wiki get locked down at the close of funding or will we leave it to its own devices, becoming an online-free-for all?' (Terras 2010).[26] It also raises a lot of new ethical questions for social scientists and humanists to grapple with. As *Nature* argues,

> For a certain sort of social scientist, the traffic patterns of millions of e-mails look like manna from heaven. Such data sets allow them to map formal and informal networks and pecking orders, to see how interactions affect an organization's function, and to watch these elements evolve over time. They are emblematic of the vast amounts of structured information opening up new ways to study communities and societies. Such research could provide much-needed insight into some of the most pressing issues of our day, from the functioning of religious fundamentalism to the way behaviour influences epidemics... But for such research to flourish, it must engender that which it seeks to describe... Any data on human subjects inevitably raise privacy issues, and the real risks of abuse of such data are difficult to quantify (*Nature* 2007).

Indeed, for Latour (2010), 'sociology has been obsessed by the goal of becoming a quantitative science. Yet it has never been able to reach this goal because of what it has defined as being quantifiable within the social domain...' so, he adds, '[i]t is indeed striking that at this very moment, the fast expanding fields of "data visualisation", "computational social science" or "biological networks" are tracing, before our eyes, just the sort of data' that sociologists such as Gabriel Tarde, at the turn of the 20th Century, could merely speculate about (Latour 2010: 116).

Further, it is not merely the quantification of research which was traditionally qualitative that is offered with these approaches, rather, as Unsworth argues, we should think of these computational 'tools as offering provocations, surfacing evidence, suggesting patterns and structures, or adumbrating trends' (Unsworth, quoted in Clement *et al.* 2008). For example, the methods of 'cultural analytics' make it possible through the use of quantitative computational techniques to understand and follow large-scale cultural, social and political processes for research projects – that is, massive amounts of literary or visual data analysis (see Manovich and Douglas 2009). This is a distinction that Moretti (2007) referred to as *distant* versus *close* readings of texts. As he points out, the traditional humanities focuses on a 'minimal fraction of the literary field',

> A canon of two hundred novels, for instance, sounds very large for nineteenth-century Britain (and is much larger than the current one), but is still less than one per cent of the novels that were actually published: twenty thousand, thirty, more, no one really

knows—and close reading won't help here, a novel a day every day of the year would take a century or so... And it's not even a matter of time, but of method: a field this large cannot be understood by stitching together separate bits of knowledge about individual cases, because it isn't a sum of individual cases: it's a collective system, that should be grasped as such, as a whole (Moretti 2007: 3–4).

It is difficult for the traditional arts, humanities and social sciences to completely ignore the large-scale digitalisation effort going on around them, particularly when large quantities of research money are available to create archives, tools and methods in the digital humanities and computational social sciences. However, less understood is the way in which the creation of digital archives are deeply computational in structure *and* content, because the computational logic is entangled with the digital representations of physical objects, texts and 'born digital' artefacts. Computational techniques are not merely an instrument wielded by traditional methods; rather they have profound effects on all aspects of the disciplines. Not only do they introduce new methods, which tend to focus on the identification of novel patterns in the data as against the principle of narrative and understanding, they also allow the modularisation and recombination of disciplines within the university itself. Computational approaches facilitate disciplinary hybridity that leads to a post-disciplinary university that can be deeply unsettling to traditional academic knowledge. Software allows for new ways of reading and writing, for example in Tanya Clement's distant reading of Gertrude Stein's *The Making of Americans* on which she writes,

> The Making of Americans was criticized by [those] like Malcolm Cowley who said Stein's "experiments in grammar" made this novel "one of the hardest books to read from beginning to end that has ever been published."... The highly repetitive nature of the text, comprising almost 900 pages and 3174 paragraphs with only approximately 5,000 unique words, makes keeping tracks of lists of repetitive elements unmanageable and ultimately incomprehensible... [However] text mining allowed me to use statistical methods to chart repetition across thousands of paragraphs...facilitated my ability to read the results by allowing me to sort those results in different ways and view them within the context of the text. As a result, by visualizing clustered patterns across the text's 900 pages of repetitions... This discovery provides a new key for reading the text as a circular text with two corresponding halves, which substantiates and extends the critical perspective that *Making* is neither inchoate nor chaotic, but

a highly systematic and controlled text. This perspective will change how scholars read and teach The Making of Americans (Clement, quoted in Clement *et al.*, 2008).

I wouldn't want to overplay the distinction between patterns and narrative as differing modes of analysis, indeed, patterns implicitly require narrative in order to be understood, and it can be argued that code itself consists of a narrative form that allows databases, collections and archives to function at all. Nonetheless, pattern and narrative are useful analytic terms that enable us to see the way in which the computational turn is changing the nature of knowledge in the university and with it the kind of computational subject that the university is beginning to produce. As Bruce Sterling argues,

> 'Humanistic heavy iron': it's taken a long time for the humanities to get into super computing, and into massive database management. They are really starting to get there now. You are going to get into a situation where even English professors are able to study every word ever written about, or for, or because of, Charles Dickens or Elizabeth Barrett Browning. That's just a different way to approach the literary corpus. I think there is a lot of potential there (Sterling 2010).

Indeed, there is a cultural dimension to this process and as we become more used to computational visualisations, we will expect to see them and use them with confidence and fluency. As we shall see later, the computational subject is a key requirement for a data-centric age, certainly when we begin to look at case studies that demonstrate how important a computational comportment can be in order to perform certain forms of public and private activities in a world that is increasingly pervaded by computational devices. In short, *Bildung* is still a key idea in the digital university, not as a subject trained in a vocational fashion to perform instrumental labour, nor as a subject skilled in a national literary culture, but rather as subject that can unify the information that society is now producing at increasing rates, and which understands new methods and practices of critical reading (code, data visualisation, patterns, narrative) and is subject to new methods of pedagogy to facilitate it. This is a subject that is highly computationally communicative and able to access, process and visualise information and results quickly and effectively. At all levels of society, people will increasingly have to turn data and information into usable computational forms in order to understand it at all. For example, one could imagine a form of *computational* journalism that enables the public sphere function of the media to make sense of the large amount of data which governments, amongst others, are generating, perhaps through increasing use of 'charticles', or journalistic articles that combine text, image, video, computational applications and interactivity (Stickney 2008). Consider the vast amounts of data that WIKILEAKS alone has generated. This is a form of 'networked' journalism that 'becomes a non-linear, multi- dimensional process' (Beckett 2008: 65). Additionally, for people in everyday life who need the skills that enable them to negotiate an increasingly computational field – one need only think of the amount of data in regard to managing personal money, music, film, text, news, email, pensions, etc. – there will be calls for new skills of financial and technical literacy, or more generally a *computational literacy*.

As the advantages of the computational approach to research (and teaching) becomes persuasive to the positive sciences, whether history, biology, literature or any other discipline, their ontological notion of the entities they study begins to be transformed and they become focussed on the computational form of the entities in their work. Here, following Heidegger, I want to argue that there remains a location for the possibility of philosophy to explicitly question the ontological understanding of what the computational is in regard to the positive sciences. Computationality might then be understood as an ontotheology, creating a new ontological 'epoch' as a new historical constellation of intelligibility.

With the notion of ontotheology, Heidegger is following Kant's argument that intelligibility is a process of filtering and organising a complex overwhelming world by the use of 'categories', Kant's 'discursivity thesis'. Heidegger historicises Kant's cognitive categories arguing that there is 'succession of changing historical ontotheologies that make up the "core" of the metaphysical tradition. These ontotheologies establish "the truth concerning entities as such and as a whole", in other words, they tell us both what and how entities are – establishing both their essence and their existence' (Thomson 2009: 149–50). Metaphysics, grasped ontotheologically, 'temporarily secures the intelligible order' by understanding it 'ontologically', from the inside out, and 'theologically' from the outside in, which allows the formation of an epoch, a 'historical constellation of intelligibility which is unified around its ontotheological understanding of the being of entities' (Thomson 2009: 150). As Thomson argues:

> The positive sciences all study classes of entities... Heidegger... [therefore] refers to the positive sciences as "ontic sciences." Philosophy,

on the other hand, studies the being of those classes of entities, making philosophy an "ontological science" or, more grandly, a "science of being" (Thomson 2003: 529).

Philosophy, as a field of inquiry, one might argue, should have its 'eye on the whole', and it is this focus on 'the landscape as a whole' which distinguishes the philosophical enterprise and which can be extremely useful in trying to understand these ontotheological developments (Sellars 1962: 36). If code and software is to become an object of research for the humanities and social sciences, including philosophy, we will need to grasp both the ontic and ontological dimensions of computer code. Broadly speaking, then, this book takes a philosophical approach to the subject of computer code, paying attention to the broader aspects of code and software, and connecting them to the materiality of this growing digital world. With this in mind, we now turn to the question of code itself and the ways in which it serves as a condition of possibility for the many computational forms that we experience in contemporary culture and society.

2
What Is Code?

In this chapter, I want to consider in detail the problem we are confronted with immediately in trying to study computer code. The perl poem, *Listen*, shown below, demonstrates some of the immediate problems posed by an object that is at once both literary and machinic (Hopkins n.d.). *Source code* is the textual form of programming code that is edited by computer programmers. The first difficultly of understanding code, then, is in the interpretation of code as a textual artefact. It forms the first part of the development process which is written on the computer and details the functions and processes that a computer is to follow in order to achieve a particular computational goal. This is then compiled to produce *executable code* that the computer can understand and run. The second difficulty is studying something in process, as it executes or 'runs' on a computer, and so the poem *Listen* has a second articulation as a running program distinct from the textual form.

The textual is the literary side of computer source code, and the example given below shows us the importance of *reading* as part of the practices of understanding source code. Even though this is a particular example of code which makes our reading seemingly easier by its poetic form, it is important to note that programmers have very specific and formal syntactical rules that guide the layout, that is the writing, of code, a style that was noted in the memorable phrase 'literate programming' (Black 2002: 131–7). As Donald Knuth explained in his book *Literate Programming* published in 1992:

> The practitioner of literate programming can be regarded as an essayist, whose main concern is with exposition and excellence of style. Such an author, with thesaurus in hand, chooses the names

```perl
#!/usr/bin/perl

APPEAL:

listen (please, please);

    open yourself, wide;
        join (you, me),
    connect (us, together),

tell me.

do something if distressed;

    @dawn, dance;
    @evening, sing;
    read (books,$poems,stories) until peaceful;
    study if able;

    write me if-you-please;

sort your feelings, reset goals, seek (friends, family, anyone);

    do*not*die (like this)
    if sin abounds;

keys (hidden), open (locks, doors), tell secrets;
    do not, I-beg-you, close them, yet.

        accept (yourself, changes),
        bind (grief, despair);

    require truth, goodness if-you-will, each moment;

select (always), length(of-days)

# listen (a perl poem)
# Sharon Hopkins
# rev. June 19, 1995
```

Figure 2.1 'Listen' by Sharon Hopkins (quoted in Black 2002: 141–2)[1]

of variables carefully and explains what each variable means. He or she strives for a program that is comprehensible because its concepts have been introduced in an order that is best for human understanding, using a mixture of formal and informal methods that nicely reinforce each other (Knuth, quoted in Black 2002: 131).

Knuth is also pointing towards the aesthetic dimension of coding that strives for elegance and readability of the final code – 'good' code. This is nicely demonstrated by The Alliance for Code Excellence, which argues for '[a] world where software runs cleanly and correctly as it simplifies, enhances and enriches our day everyday life is achievable' (ACE n.d.). Rather like the indulgences sold by the Catholic Church to pay for sins, and which led Martin Luther to break with the Church and nail his 95 theses onto the church door,[2] ACE sells 'bad code offsets' which can be used in a similar way to the use of carbon offsets.³ Where carbon offsets are means of purchasing the planting of trees or the sequestration of carbon to make up for air-flights or other carbon generating activities, code offsets allow you to program badly, but through the funding of open-source programming set aside these 'bad' practices. These offsets are not only a means of drawing attention to a real issue in any programming project where sometimes the shared norms and values of 'good code' are broken in the interests of hacking a fix or helping to ensure a product ships. They also demonstrate the way in which programmers understand their coding project, using Source Lines of Code (SLOC) as a measure of the size of a project and also draw attention to the increased likelihood of errors from 'bad code'. This online group allows programmers to purchase 'Bad Code Offsets' which,

> provides a convenient and rational approach for balancing out the bad code we all have created at one time or another throughout our lifetime—even when we can't go back and fix it directly. Denominated in Source Lines of Code (SLOC), every purchase will offset the desired quantity of SLOC and pave the way toward future code excellence (ACE n.d.).

Bad code is described as arising 'for many reasons: lack of skill, insufficient time, abject neglect or poorly documented requirements for example', further they argue that '[b]ad code weakens the utility delivered by these applications causing business loss, user dissatisfaction, accidents, disasters and, in general, sucks limited resources towards responding to the after effects of bad code rather than toward the common good' (ACE n.d.). Rather fittingly, the money raised is used to pay out 'Good Code Grants' to the open source movement to encourage more open source software.

Here it is important to draw an analytical distinction between 'code' and 'software'. Throughout this book I shall be using code to refer to the textual and social practices of source code writing, testing and distribution. That is, specifically concerned with code as a textual source code instantiated in particular modular, atomic, computer-programming languages as the object of analysis, which I will later discuss as 'delegated code'. As Cramer glosses:

> In computer programming and computer science, "code" is often understood as with a synonym of computer programming language or as a text written in such a language... The translation that

occurs when a text in a programming language gets compiled into machine instructions is not an encoding... because the process is not one-to-one reversible. That is why proprietary software companies can keep their source "code" secret. It is likely that the computer cultural understanding of "code" is historically derived from the name of the first high-level computer programming language, "Short code" from 1950 (Cramer, quoted in Fuller 2008: 172).

In distinction, I would like to use 'software' to include commercial products and proprietary applications, such as operating systems, applications or fixed products of code such as Photoshop, Word and Excel, which I also call 'prescriptive code'. In this sense 'software engineering' is the engineering, optimisation and analysis of code in order to produce software (as running or executable code). Software is therefore 'not only "code" but a symbolic form of writing involving cultural practices of its employment and appropriation' (Fuller 2008: 173). This further allows us to think about software 'hacking', which is the changing of code's function by the application of patches, software fixes or edits, as the transformation of software back into code for the purposes of changing its normal execution or subverting its intended (prescribed) functions. As an analogy we can think of code as the 'internal' form and software as the 'external' form of applications and software systems. Or to put it slightly different, code implies a close reading of technical systems and software implies a form of distant reading. This also complements Manovich's notion of 'cultural software' as those applications used to produce design, music or artistic cultural objects (see Manovich 2008). Perhaps the most important point of this distinction is to note that code and software are two sides of the same coin, code is the static textual form of software, and software is the processual operating form. This distinction, however, remains analytical, as the actual distinction between them is much fuzzier than may appear on the surface, some code is directly executable and editable *in situ*, so called interpreted code, because it does not need to go through a compilation process, and some software is self-writing, able to rewrite its functionality on the move, whether through genetic algorithms, viral coding structures or merely connective or artificial intelligence like encoding behaviour or expert systems (see the discussion of Google in Chapter 1).

These concepts, however, are not very useful unless we are attentive to the materiality of code. Getting at the materiality of code has to take into account its physicality and obduracy, but also the 'code work' and 'software work' that goes into making and maintaining the code (e.g. documentation, tests, installers, etc.), the networks and relationships, and the work that goes into the final shipping product or service. Additionally, we have to be alert to following the code's genealogy to see how it is developed as an historical object and its influences on attitudes, movements and ideas. But also thinking about code as differently and multiply articulated – both within the machine and amongst programmers and users. We must also not be afraid of using other technical devices to mediate our access to the code, indeed, even as Fuller problematises the reading of 'subscopic' code (Fuller 2003: 43), we can use devices to open any level of code to view (e.g. debuggers, IDEs, oscilloscopes). This is similar to the use by physical scientists who increasingly use technical devices to re-present the large and the small to our human dimensions through software. Naturally, the use of software to view software is an interesting methodological recursion, but with reflexivity it seems to me a perfectly reasonable way of developing tools for us to understand software and code within the humanities and social sciences.

Code

Code is a general term for a wide variety of different concrete programming languages and associated practices. When we want to look at the code, we see a number of different perspectives and scales depending on what kind of code we are viewing (assembler, C++, Pascal), on its state (source, compiled, disassembled), location (embedded, system, application) or its form (textual, visual, mapped as a graph). Further, code may also be distinguished between dominant/hegemonic code and subaltern or critical code.

Code is striking in its ability to act as both an actor performing actions upon data, and as a vessel, holding data within its boundaries. Some theorists, such as Mackenzie (2006: 5–6), have argued that for this reason it is only possibly to speak in terms of particular code. However, here I wish to argue, following Marx's approach to labour, that we can look at abstract code as a real abstraction that allows us to consider the general properties shared between different code forms. In a Clausewitzian sense, I want to think through the notion of 'absolute' code, rather than 'real' code, where real code would reflect the contingency and uncertainty of programming in particular locations. Absolute code would then be thinking in terms of both the grammar of code itself, but also trying to capture how programmers think algorithmically, or better, computationally. But to do this, it is important that we ground the discussion in terms of the particularity of examples of code that demonstrate the abstract concept's efficacy. This is how programmers themselves conceptualise code, both in terms of the abstract machine which they use to conceptualise the design and general schematics of the software/code but also in particular concrete programming practice when it comes to writing each of the functional units that make up the whole. Programmers have tended to need to continually use this hermeneutic circle of understanding 'the parts', 'the whole' and the 'parts in terms of the whole' in order to understand and write code (and communicate with each other), and this is striking in its similarity to literary creativity, to which it is sometimes compared.[4] As Larry Page (creator of Perl, a computer programming language) said, 'a language is a wonderful playground...' (Page, quoted in Biancuzzi and Warden 2009: 378). Clearly, the choice of language that programmers code in is of vital importance as different languages are expressive and functional in different ways which can help or hinder the development of a software programme. Robin Milner (creator of ML, a programming language) comments,

> Faced with a particular task, I think a programmer often picks the language that makes explicit the aspects of the task that he considers most important. However, some languages achieve more: they actually influence the way that the programmer thinks about the task. Object oriented languages have done very well from that viewpoint, because the notion of object helps to clarify thought in a remarkable variety of applications (Milner, quoted in Biancuzzi and Warden 2009: 213)

Object oriented techniques, such as object oriented design (OOD), have certainly contributed to changing the way people think about software, but when people undertake OOD it is in the sense of absolute code. Indeed, it is taken that 'absolute' code is still performative, operative and therefore, in some sense, 'runs' to the extent that it is intended to undertake or defer an action. For example, programmers sometimes use a system specification or formal language to 'run' through the programme and therefore test their ideas. The way in which it is 'run' and the extent to which this is contested by various actors associated with code will be explored throughout the book, but suffice to say that I want to highlight this performative dimension – code acts, fixes data, controls devices and communicates to other actors, and acts as a space for various forms of practices to take place. But it does not do so without limits, as Larry Page comments,

> There are the equivalent of grammar school teachers for computer languages, and for certain kinds of utterances, you should follow the rules unless you know why you are breaking them. All that being said, computer languages also have to be understandable to computers. That imposes additional constraints. In particular, we can't just use a natural language for that, because in most cases… we are assuming an extreme intelligence on the hearing end who will in turn assume an extreme intelligence on the speaking end. If you expect such intelligence from a computer, then you'll be sorely disappointed because we don't know how to program computers to do that yet (Page, quoted in Biancuzzi and Warden 2009: 382).

So, although an expressive medium, computer languages remain constrained in what may be 'said' due to the requirements that the computer in the final instance understands it. This also encourages the kind of syntactic terseness, the obscure punctuation and the layout and structure of computer programmes. Also computer programming can be an intensely social activity in addition to the perceived loneliness of the computer programmer. As Bjarne Stroustrup (creator of C++, another programming language) notes,

> A successful [programming] language develops a community: the community shares techniques, tools, and libraries… any new language must somehow manage the centrifugal forces in a large community, or suffer pretty severe consequences. A general-purpose language needs the input from and approval of several communities, such as, industrial programmers, educators, academic researchers, industrial researchers, and the open source community… the real problem is to balance the various needs to create a larger and more varied community (Stroustrup, quoted in Biancuzzi and Warden 2009: 15–16).

This contributes to the challenge of investigating code as an empirical object of analysis, whilst it is also part of a complex set of elite practices that partially forms part of the definition of the code itself (e.g. hacking, programming, etc.). As David Parnas argues, 'technology is the black magic of our time. Engineers are seen as wizards; their knowledge of arcane rituals and obscure terminology seems to endow them with an understanding not shared by the laity' (Parnas, quoted in Weiner 1994: ix). Code is therefore technical and social, and material and symbolic simultaneously. This is not a new problem but it does make code difficult to investigate and study, and similar to understanding industrial production as Marx explained, 'right down to the eighteenth century, the different trades were called "mysteries", into whose secrets none but those initiated into the profession and their practical experience could penetrate' (Marx 2004: 616). So Marx had to undertake difficult and laborious analysis of machinery, for example, before he was able to see clearly how it functioned and related to industrial capitalism.

Similarly, when understanding code there remains these difficult 'mysteries' and we must place them in their social formation if we are to understand how code and code-work are undertaken. However, this difficulty means that we also cannot stay at the level of the screen, so-called *screen essentialism*, what Waldrip-Fruin (2009:3) calls 'output-focused approaches', nor at the level of information theory, where the analysis focuses on the way information is moved between different points disembedded from its material carrier. Rather, code needs to be approached in its multiplicity, that is, as a literature, a mechanism, a spatial form (organisation), and as a repository of social norms, values, patterns and processes. As Wardrip-Fruin writes:

> Just as when opening the back of a watch from the 1970s one might see a distinctive Swiss mechanism or Japanese quartz assembly, so the shapes of computational processes are distinctive—and connected to histories, economies, and schools of thought. Further, because digital media's processes often engage subjects more complex than time-keeping (such as human language and motivation), they can be seen as "operationalized" models of these subjects, expressing a position through their shapes and workings (Waldrip-Fruin 2009:4).

This is a very useful way of thinking about code and draws attention to the way in which code, and the practices associated with it, are constantly evolving as new technologies are developed and introduced. We no longer program computers with a soldering iron, nor with punch-cards, equally, today it is very rare for a programmer to start with a completely blank canvas when writing code. Due to improvements over the last forty years or so, programmers can now take advantage of tools and modular systems that have been introduced into programming through the mass engineering techniques of Fordism. This means that software is written using other software packages, help is provided through software support programs and modular mass-produced libraries of code. In the same way that studying the mechanical and industrial machinery of the last century can tell us a lot about the organisation of factories, geographic movements, materials, industries, and processes in industrial capitalism; through the study of code we can learn a lot about the structure and processes of our post-Fordist societies through the way in which certain social formations are actualised through crystallisation in computer code. This is certainly one of the promises of software studies and related approaches to understanding software and computational devices.

It is important to remember, however, that code nonetheless 'exists' within the virtual space of a digital computer, that which Castells (1996) called the 'space of flows'. That is not to say that code does not also exist in paper documents, schemata, files and folders, and the mind of the programmer, which of course it does. However, code 'work' is written inside the computer within the programming editing software, and code is compiled and run on the computer too. This means that software is mediating the relationship with code and the writing and running of it. When we study code we need to be attentive to this double mediation and be prepared to include a wider variety of materials in our analysis of code than just the textual files that have traditionally been understood as code. For example, many development environments now allow the programmer to create visual interfaces in 'interface builders' that are akin to image-drawing software. These produce collections of files (e.g. 'nib' files) that contain the logic of the interface but files that are not code in the normal sense of the term, yet are crucial to the compiling and execution of the software. In a technical sense these files contain the interface objects and their relationships in a saved format through a process called streaming which serializes the objects into a file format, called 'freeze dried', as the object are literally frozen in their current state and saved to disk.

Code's relationship to the real world is indirect, itself mediated through frames and models that attempt to capture some aspect of the real world and present it to the code for analysis. Mapping this distantiated code and the logic and processes that it follows, not to mention the textual and processual forms that structure it, are therefore extremely challenging. We have an object of research that is in danger at the moment of its capture of being perceived as that which has been frozen and turned into a feeble simulacra of itself, whether as textual or screenic source or interface graphics. Manovich (2008: 17) argues that because of the mediation of software we should talk about the 'media interface' to highlight that we are very rarely using the media objects as things-in-themselves. Rather they are available only as 'media performances' that can only be understood through a notion of software studies. Media performances, refers to the fact that increasingly our media is constructed on the fly from a number of modular components derived from a wide variety of networked sources. Code is therefore connective, mediating and constructing our media experiences in real-time as software. Code must then be understood in context, as something that is in someway potentially running for it to *be* code. Code is processual, and keeping in mind its execution and agentic form is crucial to understanding the way in which it is able to both structure the world and continue to act upon it. Understanding code requires a continued sensitivity to its changing flow through the hardware of the technology. Indeed, this is as important as placing code within its social and technical milieu or paying attention to the historical genealogy.[5]

This, perhaps, gives us our first entry point into an understanding of code; it is a declarative and comparative mechanism that 'ticks' through each statement one at-a-time (multiple threaded code is an illusion except where several processors are running in a machine simultaneously, but even here the code is running sequentially in a mechanical fashion on each processor). This means that code runs sequentially (even in parallel systems the separate threads run in an extremely pedestrian fashion), and therefore its repetitions and ordering processes are uniquely laid out as stepping stones for us to follow through the code, but in action it can run millions of times faster than we can think – and in doing so introduce qualitative changes that we may miss if we focus only on the level of the textual.

Although the speed at which computers work may seem unbelievable, it is interesting to note that if the speeds of the computers were slowed down, we would be able to watch our computers 'tick' through their actions in real time. Indeed, this is exactly what software called 'debuggers' enables, by allowing the programmer to follow the code in detail. By slowing down or even forcing the program to execute step-by-step under the control of the programmer the branches, loops and statements can be followed each in turn in order to ensure the code functions as desired. In some sense then, the quantitative speed of computer processing gives rise to a qualitative experience of computers as miraculous devices.

Code is also a relay through which action is carried out, but for this to be achieved, the external 'real' world must be standardised and unified in a formal manner which the code is able to process and generate new data and information – and this we can trace.[6] This is where a phenomenology of code, or more accurately a phenomenology of computation, allows us to understand and explore the ways in which code is able to structure experience in concrete ways. By following the code and its textuality and structure, we can focus on the *pragmata* of code and hence on its materiality.

The second entry point into understanding code is that computer code is manufactured and this points us towards the importance of a political economy of software. Software is not written by machines, but rather by human beings, often one line of code at a time.[7] It is a slow, time-consuming and often painful activity that is full of mistakes, trial-and-error testing, etc. in implementation. As Yukihiro Matsumoto explains,

> Most programs are not write-once. They are reworked and rewritten again and again in their lives. Bugs must be debugged. Changing requirements and the need for increased functionality mean the program itself may be modified on an ongoing basis. During this process, human beings must be able to read and understand the original code; it is therefore more important by far for humans to be able to understand the program than it is for the computer (Oram and Wilson 2007: 478).

Code is labour crystallised in a software form that is highly flexible and which when captured may be executed indefinitely. Code therefore operates as a continual process, and 'the main point is that every successful piece of software has an extended life in which it is worked on by a succession of programmers and designers...' (Bjarne Stroustrup, quoted in Oram and Wilson 2007). This is not to say that software cannot be inflexible. Code has to be very carefully coded in such a way as to write in the ability to be flexible, or forgiving, otherwise it is liable to run incorrectly, or even corrupt data stores and outputs due to the problems encountered with poorly formatted input data. However, code needs to be thought of as an essentially unfinished project, a continually updated, edited and reconstructed piece of machinery.

Code is also treated as a form of property in as much as it is subject to intellectual property rights (IPRs), like copyright.[8] Owning code can therefore be a very lucrative activity and owning the copyrights a key part of what constructs a market in software such as Microsoft Windows. But code differs from a factory or machine (which in a certain sense can be considered condensed physical labour) in that it is the processes of thought itself that is being transferred into software. The thinking actions of the programmer (and sometimes the tacit knowledge of the workers whose skills are being encoded) are abstracted into the programming language (sometimes through the absorption of the tacit knowledge of experts) and then encoded (stabilised) within software. This is what Hardt and Negri (2000) named 'immaterial labour', pointing to the way in which contemporary capital increasingly requires that our intellectual labour is alienated into machines. However, that is not to ignore the attempts by owners and managers to move software development from a craft-like method to an industrial processing model of creating software by using Taylorist techniques, like time-and-motion studies, peer review programming, software libraries, modularity, and so forth. This has certainly transformed the process of writing of the majority of software into something approaching a Fordist way of producing software. This can be understood as a move from literary code to an engineering of industrial code. However, for the creation of specialist software, particularly for time-critical or safety-critical industries, the literary craftsmanship of programming remains a specialised hand-coded enclave.

It is interesting to think about the way in which today code is written through a process of collage, whereby different fragments of code (usually called '#includes') are glued together to form the final software product (this is a key principle behind software libraries and object oriented programming). This naturally undermines the notion of a single author of software, for instance, programmers themselves do not 'reinvent the wheel' and instead reuse old, reliable code wherever possible. It also highlights that running code is a collective achievement. Somehow then, we must keep in mind both the 'code work' that takes place in producing and maintaining software, but also its extremely important social and sharing dimensions. This also must be connected to the notion of supporting institutions and the key technical assemblage that is required to keep programmers programming, such as technical facilities, libraries, furniture, light, heat and a salary. In summary then, we must keep the political economy of code present in our minds as we consider the specificity of codes making and running, even if, as in this book, it is backgrounded whilst we focus on the phenomenology of code.

Thirdly, it is important to note that software breaks down, continually. In some of the more imaginative claims made by proponents of the Information Society it is often forgotten the difficult work of making software function with other software (whether or not it is hidden within the black-box of a particular technical device). Adding software to a system may make it cheaper, easier to change, or even quicker, but it does not make the system more reliable (Weiner 1994: xv). One of the most arresting demonstrations of software breakdowns concerns a combat operations centre in the US:

> Constructed in 1961, the U.S. Air Force's underground combat operation center inside Cheyenne Mountain, Colorado, experienced alarming software failures. For eight tense minutes in 1979, Cheyenne mistook a test scenario for an actual missile attack, a mistake that could have triggered a nuclear holocaust (Hughes 2005: 90).

In this case catastrophe was avoided, but today software is even more embedded and implicated in running many more systems that interconnect in ways that are difficult to manage and understand. Our current knowledge and capabilities with regard to software are extremely immature, indeed, sensitive or time-critical code production is still produced as a craft-like process (usually with a wrapper of management discourse disguising it). The implications are interesting; much software written today never reaches a working state, indeed a great quantity of it remains hidden unused or little understood within the code repositories of large corporate organisations.

These code repositories also tend to include a detail history of 'commits', that is, the way in which the software and documentation changed over time. Commits are contributions to the codebase that are slowly rolled into a stable release after testing. Before the stable release software tends to go through alpha and beta releases, these used to be private versions but increasingly beta is seen as a way to encourage user feedback and testing by early release, nonetheless, it is expected that the stable release has ironed out the most serious errors in software. However, small bugs can remain in the code and there can be subtle effects from a simple programming error which can have a potentially catastrophic effect on a major corporation, for example,

> On Wednesday, November 20, 1985, a bug cost the Bank of New York $10 million when the software used to track government securities transactions from the Federal Reserve suddenly began to write new information on top of old… The Fed debited the bank for each transaction, but the Bank of New York could not tell who owed it how much for which securities. After ninety minutes they managed to shut off the spigot of incoming transactions by which time the Bank of New York

owed the Federal Reserve $65 billion it could not collect from others... Pledging all its assets as collateral, the Bank of New York borrowed £46.8 billion from the Fed overnight and paid £10 million in interest for the privilege. By Friday, the database was restored, but the bank... lost the confidence of investors (Weiner 1994: 11; using 2010 dollar values).

Although code is always vulnerable to disruptions and failure through *internal* contradictions, it is also threatened by the intrusion of other *external* disruptions and failures, for example from institutional orders (such as a corporate take-over or when the software project is cancelled) or in the resistance of its users and programmers (both internal or external) and from rival projects and products. Software, therefore, always has the possibility of instability being generated from within its different logics. Even when it is functioning, code is not always 'rational'. There is always the possibility for unintended consequences that come from misunderstood or unexpected scenarios, errors, bugs and code is as affected by the passage of time as much as any other material artefact.

Therefore, and lastly, software, contrary to common misconceptions, follows a cycle of life. It is not eternal, nor could it be. It is created as code, it is developed whilst it is still productive, and slowly it grows old and decays, what we might call the *moral depreciation* of code (Marx 2004: 528).[9] In software, this is usually recognised by the upgrading of software, for example through software updates, or through the development of new methods or processes of software design, for example from procedural to object-oriented programming.

Hence, we must be conscious of the fact that software ages.[10] And often ages very badly. Like glass, code crystallises and ages at different rates, some code becoming obsolete before others, making replacement increasingly difficult or incompatible. Computers grow old and weary, the circuit-boards and processors become difficult to replace, the operating system is superseded and the development tools and skills of the programmers become obsolete. Indeed, in one large multinational company that I worked for, I was always fascinated by a beautiful bright orange PDP-11 computer that was used to run an important (and extremely profitable) financial system which could not be replaced for many years. The program originally written in PDP-11 assembly language had long been forgotten and the programmers had left the company or retired – in fact the code had long been lost because of the tendency to 'patch' the software executable rather than rewrite the code. Additionally, the manufacturer, DEC, no longer supported the computer model, nor was inclined to, and few if any employees had a desire to learn a computer system and associated programming language that was by definition useless for their careers. In the end, after many decades, it was replaced and then, finally, turned off. So software too can suffer a kind of death, its traces only found in discarded diskette, the memories of the retired programmers, their notebooks, and personal collections of code printouts and manuals.[11] And strangely, there is so far little attempt to build museum collections and store these passing memories of long-past software as cultural knowledge in museums or library collections, although the Computer History Museum, in California in the US, and the National Media Museum, in Bradford in the UK, are notable exceptions.[12]

This complexity adds to the difficulty of understanding code, as Minsky observes,

> When a program grows in power by an evolution of partially understood patches and fixes, the programmer begins to lose track of internal details, loses the ability to predict what will happen, begins to hope instead to know, and watches the results as though the program were an individual whose range of behaviour is uncertain... This is already true in some big programs... it will soon be much more acute... large heuristic programs will be developed and modified by several programmers, each testing them on different examples from different [remotely located computer] consoles and inserting advice independently. The program will grow in effectiveness, but no one of the programmers will understand it all. (Of course, this won't always be successful – the interactions might make it get worse, and no one might be able to fix it again!) Now we see the real trouble with statements like 'it only does what the programmer told it to.' There isn't any one programmer (Minsky quoted in Weizenbaum 1984: 235).

The ontology of Code

These discussions have highlighted the importance of an interdisciplinary range of methodologies and approaches to understanding code, and certainly make the idea of a single approach extremely problematic. Nonetheless we should be clear that the ontology of code is specifiable, indeed, programmers already know what code is, *qua* code. For example, each of the following highlights the way in which an ontology of code is inculcated in the programmer and serves to reinforce how code is both understood and materialised as part of programming practices.

Through habituation/training/education

Programmers are taught from an early age to recognise what is and what is not code. This takes place both in moments of experimentation at the computer, but also in training programmes, education and so forth. These methods of habit become deeply ingrained in the way in which a programmer will tackle a computing problem and are demonstrated by the way in which programmers often become attached to particular programming languages, shells, editing environments or computer platforms (e.g. Unix). Programmers are taught relatively self-contained abstract problems to solve on computer science degrees usually in a limited range of programming environments, whereas when they enter formal work they often enter a new phase of training involving complex system interdependencies and legacy systems. Both of these initial experiences tend to encourage the reliance on tried and trusted platforms and solutions, to which the common refrain when deciding on a computer platform is 'no-one has ever been fired for buying Microsoft Windows'. This has been reinforced by marketing efforts which use 'fear, uncertainty, and doubt' (FUD) to encourage customer loyalty.[13]

Through structural constraints (e.g. IDE, compiler)

In addition to the habituation and education of programmers are the constraints offered by the programming environments themselves which can be very unforgiving. Punctuation, for example, is part of the syntax of programming languages and misplaced punctuation can cause all sorts of strange bugs and errors to occur. In 1962, an Atlas-Agena rocket had an incorrect equation in its computerised guidance system was carrying Mariner I, a space exploration worth $18.5 million, into space. Unfortunately the equation was missing a 'bar' – a horizontal stroke over a symbol that meant that the computer should use a set of averaged values, instead of raw data. This caused a miscalculation in the navigation computer which reported that the rocket was behaving erratically and so attempted to 'correct' the error, which actually caused erratic behaviour. The controllers were therefore forced to blow up the rocket to safeguard the community at Cocoa Beach (Weiner 1994: 4–5).[14] Code therefore requires a very high degree of proof-reading ability in order to ensure the correctness of the program under development. Where the errors are egregious the compiler will soon alert the programmer to the problem and its location, but with very subtle errors, often involving punctuation, the error might be almost undetectable, for example only manifesting itself in the actual operation of the programme under very specific conditions, these intermittent bugs are very difficult to avoid (see a fictionalised account in Ullman 2004). Hence the development environments try very hard to prescribe onto the programmer very clear structural constraints, for example through source code colouring, automatic formatting and layout, and through restrictions on the way in which a program may be developed (e.g. requiring type declarations, classes and so forth to be explicit). We can think of this as a form of prescribed literate programming.

Through a constellation of shared knowledge and practices

The way in which a programme is written is not only a private activity. The source code will likely at some stage be maintained by others, consequently, programming can be an extremely social activity with shared norms as to how the code should be laid out, what are the acceptable and unacceptable ways of writing a program. Commentary code, for example, is often used to describe the way in which a program functions, but when being shared it may still be important to use shared notions, as well as clear names for variable and function names (the Obfuscated C Code Competition discussed below is an interesting counter-example of this). Techniques such as agile programming, which encourages programming with a partner, code reviews, where a committee double-checks the logic of the code, and free/libre and open source software (FLOSS) which allows anyone to download and read the code, all act to promote readability and excellence in the programming.

Additionally we must not forget that computers, too, already know what code is, *qua* code, again through the particular materiality offered by the computer hardware that is a condition of possibility for the functioning of software:

Microprocessors have a limited vocabulary defined by their instruction set (as microcode and/or as assembly language)

For a program to execute requires that it be written in the precise form that a computer expects in order to run it. This means conformance with its programming operation, its machine language codes, and the file formats and structures that are prescribed. Usually the processor will have an instruction set specific to it which means that a binary files (usually the form the application, such as Word, is distributed) runs only on specific processors. This explains why PC software does not run on a Mac and vice versa (although with the speeds of processors becoming faster and faster, emulation of another processor is an increasing possibility for running applications).

Compilable and executable code is structured and formatted by precise rules

Computers will not compile code that is not compilable. That is, if the written code does not abide by the rules that structure the program it will not be able to translate it into a form that the computer can execute. This is of two forms: (1) Programs that are logically coherent but are programmatically incoherent, that will compile but not perform any useful action; and (2) Programs that are programmatically coherent but which are logically incoherent. In this case the compiler will be unable to translate the program into a functioning executable. In some cases a compiler can be set to loosen the boundaries of what may compile (for example ignoring deprecated functions, that is, functions that are no longer supported), but this may then result in programs that simply do not work, or not as expected.

Metaphorical code

There is also a metaphorical cultural relation to these ontologies which have become cultural tropes that structure the way in which people understand code. We can think of these as the grand narratives that are used to explain what code is and how, by analogy, it functions. This is reflected in the use we make of metaphors to think about computers. These also tells us a lot about our displaced ideas of how code works and what code does. Some of the major tropes include:

Code as an engine

One of the most common tropes used to describe computer code is the metaphor of the engine. Here the notion of the processor as the hardware device that performs a processing workload finds its analogue in the code that actually defines the task to be undertaken and run on the hardware. This idea draws its inspiration from mechanical understandings of the use of tools to undertake manual tasks and later the notion of machinary as machines that make machines. Code from the standpoint of its use as an engine has returned as a common way of discussing specialised processing platforms, such as 3D gaming engines or search engines. The trope shapes the way in which code has traditionally been seen as a mechanism and how this has influenced the way in which it has been developed and maintained both within and outside corporate and organisational boundaries. This trope focuses on code that does things. That is, code that lies within material functional processes and procedures, that monitors, controls, manufactures and directs. One of the classic inspirations for this metaphor lies with the Difference Engine created by Charles Babbage (1792–1871) who designed a machine for processing symbols. This used a number of rotating cylinders, shafts and cranks to compute values of polynomial functions but which due to the complexity and cost of the project was never actually built. He later went on to develop the Analytical Engine which, although also never completed, would have allowed programs to run through the use of a form of punch-card. Ada Lovelace, a female mathematician, who actually wrote a program to be executed on the machine is therefore widely credited within computing circles of being the first computer programmer. The Analytical Engine had a unit to perform arithmetic calculations which he called the 'Mill' together with a rudimentary memory area called 'Barrels'. The movement of symbols throughout the machine were handled by mechanical registers which conveyed the values through the system and which could be stored, calculated, and written out to a rudimentary printer, punch-cards and a bell (Beniger 1986: 399). The designs generated by Babbage were inspired by the use of cards to 'program' mechanical looms such as the Jacquard loom developed in 1801 to program woven patterns in power looms. This notion of computation through mechanical processes was further embedded in cultural representations by the use of a variety of mechanical devices. From 1880 up until the present age, mechanical methods of simplifying or automating calculations were sought (Beniger 1986: 400–401).

It was perhaps with Konrad Zuse's attempt to design a universal calculating machine in 1934 in Berlin, that began the process of movement from a purely mechanical, to an electromechanical relay machine. Whilst at the same time in the US, John Atanasoff was developing a purely electronic machine based on vacuum tubes, a shift from electromechanical to electronic completed around 1939. Howard Aiken, working at Harvard was also drafting a proposal for an electromechanical calculator called the Mark I, later completed by IBM in January 1943. Whilst there were a number of key innovations in this field, finally with theoretical contributions from Alan Turing (1936), Emil Post (1936), and Claude Shannon (1938) the notion of calculation moved from a problem of arithmetic to that of logic, and with it the notion the 'information can be treated like any other quantity and be subjected to the manipulations of a machine' (Beniger 1986: 406). This was the beginning of the use of logic to perform symbolic processing, and therefore the move towards a binary system of digital processing. It was also the move away from the metaphor of the engine and towards a notion of computation and symbolic processing.

Code as an image or picture

This is code at the level of the interface, the screenic dimension of code. This trope tends to see the screen and the interface as crucial dimensions of understanding code, and have a tendency towards screen essentialism. Nonetheless, the development of interfaces and human computer interfaces in general was a critical breakthrough that facilitated the wide-spread adoption of computer technology – and indeed is spurring the new wave of mobile devices that have to open up entirely new interfaces and representational forms (e.g. data visualisation). This trope points towards a historical contextualisation and present new forms of interface in terms of developments of key events in the history of computer science, the visual, and sometimes connects to art history. There is also an aspect to the aesthetic dimension of code. More particularly it asks questions about: (1) the way in which coding itself becomes an aesthetic pursuit (e.g. MacKenzie 2006), thought in terms of 'beautiful code', that is code that is readable, focused, testable and elegant (Heusser 2005). An example of which given by Jon Bently in (Oram and Wilson 2007) in which he describes beautiful code by saying that the rule that '"vigorous writing is concise" applies to code as well as to English, so [following this] admonition to "omit needless words"... this algorithm to sort numbers is the result':

```
void quicksort(int l, int u)
{   int i, m;
    if (l >= u) return;
    swap(l, randint(l, u));
    m = l;
    for (i = l+1; i <= u; i++)
        if (x[i] < x[l])
            swap(++m, i);
    swap(l, m);
    quicksort(l, m-1);
    quicksort(m+1, u);
}
```

Figure 2.2 An example of 'beautiful' code as a sorting algorithm (Oram and Wilson 2007: 30)

Bentley further writes:

'I once heard a master programmer praised with the phrase, "He adds function by deleting code." Antoine de Saint-Exupéry, the French writer and aviator, expressed this sentiment more generally when he said, "A designer knows he has achieved perfection not when there is nothing left to add, but when there is nothing left to take away." In software, the most beautiful code, the most beautiful functions, and the most beautiful programs are sometimes not there at all' (Oram and Wilson 2007: 29).

Code aesthetics also raises questions about (2) the way in which artists and musicians and increasingly drawing on the resources of code to create, for example, time-based artistic installations and code-based artforms (such as code poetry). For example Sharon Hopkins's perl poem 'rush':

rush by Sharon Hopkins, June 26, 1991

```
'love was'
  && 'love will be' if
  (I, ever-faithful),
    do wait, patiently;
"negative", "worldly", values disappear,
  @last, 'love triumphs';
join (hands, checkbooks),
  pop champagne-corks,
  "live happily-ever-after".
   "not so" ?
     tell me: "I listen",
       (do-not-hear);
push (rush, hurry) && die lonely if not-careful;
"I will wait."
   &wait
```

Figure 2.3 'Rush' by Sharon Hopkins (Hopkins n.d.)

Code as a medium of communication

Code here is understood as a form to facilitate communication, transformation, transfer of data and information, voice calls and all other sorts of media. From looking at Shannon and Weavers' early work on information theory and the transmission of information, to the reconfiguration of broadcast and radio spectrum in terms of networked packet switching forms, code is changing how the entire radio spectrum is regulated, assembled and used by broadcasters and users. Code allows incredible flexibility in handling communicational channels and is therefore blamed with causing disruptive innovation in terms of the mass media (sometimes theorised as Media 2.0) through its reconfiguration of the underlying physical resources, code is changing institutions, such as the BBC, for example, through the convergence on the digital form in television, radio, cinema, etc. The BBC have responded with the BBC iPlayer, a free to Internet replay mechanism using software and also their Project Canvas, a 'proposed partnership between Arqiva, the BBC, BT, C4, Five, ITV and Talk Talk to build an open internet-connected TV platform' – now called 'You View' (Project Canvas 2010). This communicational trope tends to connect to questions raised in media policy, such as the regulatory system in the nation state and arguments for a tacit particular broadcasting model. Code is therefore understood through the existing regulatory frameworks which have been used to regulate television, radio, and telephone communicational forms. The metaphor of code as a communications channel also helps to explain recent discussions about 'net neutrality' and 'search neutrality' through an understanding of code as a communications medium.

Code as a container

This form of code has been rather overlooked, perhaps because of its perceived passivity or the usual attitude towards technology that is hidden behind the interface. Here again, I want to materialise the code as container by pointing towards the modern growth in server farms, cloud computing and the like to understanding the hidden world of computer storage. Databases, collections, archives, data centres and similar inventory forms of code are crucial to the information society and without them many of the breakthroughs in contemporary technological forms (such as the iPod which stores its music in the form of a database) would not have been possible. Kirschenbaum (2004) offers an exemplary example of researching code as a container, where he undertakes a 'grammatology of the hard drive' through looking at mechanisms of extreme inscription of magnetic polarity on the hard disk platters to understand this form of 'electric writing'. This involves a micro-analysis of code-based devices, but we can also think of macro-analysis of data centres and DNS routers etc. Data centres, in particular are usually large, simple, centralised data storage centres that physically hold the large quantities of computer data. Their physicality demonstrates the materiality of computer data and networks, particularly the seemingly ephemeral 'cloud computing'. They are also repositories of extremely complicated switching systems, virtualisation platforms, that allow different operating systems to be run simultaneously, and even a form of containerisation, whereby the capacity of a data centre can be increased as load grows through physical transference of extra servers.

Most current containerised data centres, made by Hewlett-Packard, IBM, SGI and others, are built using standard 20- and 40-foot shipping containers although there are 'moves to larger sizes allow a company to add extra compute[r] capacity in less than 100 days, versus a year or more to build a new data centre. They also defer the high costs of building a new facility, and they generally can be made much more energy efficient' (Niccolai 2010).

Towards a grammar of code

To help us to think about code more analytically, in this section I would like to introduce tentative Weberian 'ideal-types' to help us think about the different forms or modalities of code, namely: (i) digital data structure, (ii) digital stream, (iii) delegated code, (iv) prescriptive code, (v) commentary code, (vi) code object and (vii) critical code. Ideal-types are an analytical construct that are abstracted from concrete examples. They also provide a means whereby concrete historical examples of code may be compared and allows us to consider the ways in which code might deviate from this form. The relatively high-level abstract ideal-types I discuss in this section are intended to help make code more clear and understandable; to develop an understanding of the kinds of ways in which code is manifested; and help to reduce ambiguity about code by providing a means to develop adequate descriptions that contribute to understanding code's historical characteristics (see Morrison 1997: 270–3).[15] Too often the question of digital media is ignored or discussed in essentialist or contradictory ways. By creating these ideal-types I aim to unpack the different modalities of code (as a digital form) and allow us to develop our understanding of the way in which it is used and performed in computer technology.

Data

In the static atomic form of digital data storage and transmission, the digital generally has a passive relationship with technology; data doesn't do anything of itself. The term *digital code* or the *digital* is often used to broadly refer to the digital collection of 0s and 1s that can be used to store functions for operating a computer (i.e. machine-code) and alternatively for storing information (i.e. binary data). Different forms of data structures are stored in the memory of the computer or hard disc in the encoding of binary data – as rows of 0s and 1s in patterns and grids. However, digital data is also the result of the discrete way in which computers, and digital technology in general, translate the analogue continuous phenomenal world into internal symbolic representational structures. These structures are limited to specific sizes that are 'fitted' to the external world by a translating device such as an analogue-digital converter. Data is therefore a key element of understanding code and as an analytic category it allows us to understand the way in which code stores values and information in a form that is stable and recallable.

Code

Computer code is involved with action, in terms of processing, and articulation, in terms of the screenic dimension, within the computer. Code is an unfolding process and performs a number of particular functions. A function, within code, is 'a self contained section of code... that is laid out in a standard way to enable deployment and re-use at any number of different points within a program' (Fuller 2008: 101). Code can be understood as the mechanism that operates upon and transforms symbolic data, whether by recombining it, performing arithmetic or binary calculation or moving data between different storage locations. As such, code is operative and produces a result (sometimes at the end of a number of sub-goals or tasks), often in an iterative process of loops and conditionals.

Delegated code (or source code)

Code has a dual existence, as delegated code residing in a human-readable frozen state that computer programmers refer to as 'source code', and as 'autonomous' prescriptive code that performs operations and processes. For example, in computer programming, to explain how a particular piece of code works, and to avoid talking about a particular instantiation of a programming language, algorithms are written out in 'pseudocode'. That is in a non-computer, non-compilable language that is computer-like but still contains enough natural language (such as English) to be readable. That is, the algorithms allow the process to be described in a platform/language independent fashion, which can be understood as a pre-delegated code form. This is then implemented in specific programming languages. But these algorithms eventually have to be turned into a computer programming language that can be compiled into prescriptive code, and therefore run as software. Computer code has a distinctive look as an often incomprehensible collection of English keywords, symbols and idiosyncratic spacing and layout. The source code itself is static and is generally written in source-code files that are text-based files, although many of the more sophisticated editors now display code in colour to help the programmer write. However, it is important to note that as human-readable text files, the delegated code is also open to interpretation by different 'readers', whether human or machine (see Marino 2006 for a discussion of this). Nonetheless, at some point the abstractions manipulated by the programmer within delegated code will have to be translated into the necessary binary operations for the 'correct' functioning of the prescriptive code.

Prescriptive code (or software)

For the computer to execute the source code as human-readable delegated code it would need to be translated into an executable, that is, machine-readable prescriptive code (see Stallman 2002: 3). This is how we tend to think of software. At machine level, executable code, as prescriptive code, is represented digitally as a stream of 0s and 1s and is very difficult for humans to write or read directly. To the human eye this would look like long streams of 0s and 1s without structure or meaning, hence they are often referred to as machine-readable files (insinuating the inability of humans to understand them directly in contrast to human-readable files). Indeed, the mythology of expert programmers and hackers dates back to the times when this was one of the only means of programming computers (Levy 2001). The production of computer code at this low level would be prohibitively complex, time-consuming and slow for all but the smallest of programs. The programmer simplifies the act of programming by abstracting the code implementation from the actual machine hardware. Prescriptive code is usually packaged and sold as a finished product, such as Microsoft Word, without the underlying source-code included in the distribution.

Critical code

This is code that is written to open up existing closed forms of proprietary computer code, providing users the ability to read/hack existing digital streams and hence to unpick the digital data structure (see below).[16] This could be where software lock-in has become a particular problem, such as with using proprietary data formats. Here, I also want to include code that is designed to hack existing closed proprietary code – often encoded as prescriptive code, such as DeCSS, which by careful examination of DVD prescriptive code opened up the DVD format for GNU/Linux users (Mackenzie 2006: 28–9). Equally, the recent Jailbreak software for unlocking the Apple iPhone and iPod Touch, hacks the prescriptive code that controls the phone with the intention of opening the hardware and software platform up to the developer/user. Critical code is drawn from the concept introduced by Fuller (2003) of 'critical software', but is contrasted to prescriptive code in the normative content of the delegated code. I am thinking particularly of free software and open source projects here such as the GNU/Linux operating system (see Berry 2004; Chopra and Dexter 2008: 37–71). Therefore, a requirement of critical code would be that the source/executable would be available for inspection to check the delegated processing that the code undertakes. If the source was unavailable then it would be impossible to check the prescriptive code to ensure it was not bogus or malicious software and it could not then be critical code.

Commentary Code

Delegated code is written in preliminary documents that contain the logic of program operation. But, in addition to the controlling logic of the delegated code program flow, the source code will often contain a commentary by the programmer in a special textual area usually delimitated by special characters (e.g. '<!--' tag in Javascript). These comments assist both the programmer and others wishing to understand the programming code and I introduce the ideal-type commentary code to describe these areas. These textual areas are used to demonstrate authorship, list collaborators and document changes – thus source code offers a hermeneutic and historical record in commentary code in addition to the processing capabilities of the explicitly delegated code within the file. When compiled into software this commentary code is usually stripped out of the source-code.

Digital data structure

This is the static form of data representation within the storage systems of a computer system. The digital encoding of analogue information (such as in the ripping of an old vinyl LP) is the transfer from one medium of storage (continuous grooves in vinyl) to another (discrete values that can represent waveforms). Something of the detail is always lost when moved from the phenomenal world to the discrete world of the computer. Digitalisation is therefore the simplification and standardisation of the external world so that it can be stored and manipulated within code. For example, music stored within the computer is translated from its analogue waveform (which is a continuous wave) and quantised into discrete 'chunks'.[17] This highlights the importance of a focus on the materiality as different embodiments fix data in different ways; think here of the different 'codecs' (coding-decoding modules) that are used to fix moving audio-video, such as MPEG, MP4, OGG and DIVX. Similarly, in any transmission, digital data is broken down to its most basic level as a string of 0s and 1s and chopped into neat packets of data and sent through a network, rather like little parcels sent through the post.

Digital stream

When computers store media content to a hard disc or other medium, the media is encoded into binary form and it is written to a binary file as a digital stream, as a one-dimensional flows of 0s and 1s. Data is transmitted across networks or through other mediums (such as radio). Within the digital stream file there are markers (such as the file-type discussed below), structural forms and data patterning that provide an encoding that allows the computer to bring the data back to its original depth as a digital data structure. To do this the computer relies on standard file and data structures to decode these binary files. When the file lies on the hard disc its functionality remains inert and static as a digital stream, for the file to become usable requires that the computer re-read the digital stream back into the computer and re-create the hierarchical structure. In a similar way, we take delivery of 2D flat-pack furniture from Ikea (the digital stream) and are required to read the instructions (the file structure) to piece together and rebuild the 3D wardrobe (the digital data structures located inside the computer memory) prior to being filled with clothes (or 'run' on the computer). The flexibility of being able to render information, whether audio-visual or textual, into this standardised digital stream form allows the incredible manipulation and transformation of information that computers facilitate (e.g. Unix uses a digital stream of text as the ubiquitous universal format in the operating system). This stream format also enables the access, storage and relational connections between vast quantities of data located in different places, such as demonstrated through search engines like Google. Eric Schmidt, Google's chief executive describes Google as '"a company that's founded around the science of measurement," and it is striving to "systematize everything"' (Carr 2008). This translational quality of digital representation and storage (albeit at an often degraded resolution within digital data structures) is something that highlights the strong remedial qualities of digital representation.

Code objects

At the humanised level of abstraction of third generation languages, delegated code can become extremely expressive and further abstraction is made easier. Away from thinking in terms of digital data structures or digital streams, the programmer begins to think in terms of everyday objects that are represented within the language structure, so rather than deal with 0s and 1s, instead she might manipulate another ideal-type which I will call code objects–such as 'cars' or 'airplanes' which have related properties such as 'gas tank' or 'wing span' and functions (i.e. methods) such as 'accelerate' or 'open undercarriage'. The further the programmer is positioned from the atomic level of 0s and 1s by the programming language, the further from the 'metal'–the electrical operation of the silicon. Therefore the programmer is not required to think in a purely mechanical/electrical fashion and is able to be more conceptual. There is a growing use of the concept of the discrete 'object' within computing. It is used as a monad containing a protected internal state, methods and interfaces to its external environment. This 'object' is used within the source code as a technique called object-oriented programming, as an abstraction where it is deployed as a conceptual metaphor for users to manipulate digital artefacts (see Scratch n.d for a visual example), and also as an active process within a network of programs, users and other objects. It allows a greater degree of modularity and automation within software and is increasingly in use at the level of the user interface.

Functions/methods

These are discrete parts of code that do things, usually the processing or iterative actions on data, particularly the digital data structure. In procedural programming languages they were called functions, and in object-oriented programming languages they are called methods. Essentially, these areas of the code can be used and reused and are usually written in a general fashion to be applicable to standard data types. Many operating systems now supply a library of handy functions/methods that are used by programmers and standardised across the platform called Application Programming Interfaces (APIs).

Web 2.0 and network code

I would like to spend a few pages thinking through the questions raised by certain new developments in Internet technologies related to Web 2.0 and social media. This is because the technologies that make up the Web 2.0 notion have been hailed by many technologists as a revolutionary new way for the Internet to function, with rich audio-visual material, interactivity, speed, efficiency and a specifically social dimension to the user experience. As Web 2.0 and its recent cousin 'Cloud Computing'[18] remain important, if somewhat nebulous terms, within the technology industry, they are important to mark when we discuss the changing shape of code and its increasing sociality.

Indeed, Web 2.0 is not a technology, as such, rather it is an ideal for the way in which certain social technologies might be imagined as working together to create useful applications. It is a technical imaginary intended to create the possibility for rethinking a particular technical problem – particularly the Internet as it existed in 2004. These aspects of the old web were usefully glossed by O'Reilly (2005a) when he attempted to outline what the major differences between Web 2.0 and preceding technologies, 'Web 2.0 is the network as platform, spanning all the connected devices; Web 2.0 applications are those that make the most of the intrinsic advantages of that platform' (O'Reilly 2005b).

Web 1.0		Web 2.0
DoubleClick	-->	Google AdSense
Ofoto	-->	Flickr
Akamai	-->	BitTorrent
mp3.com	-->	Napster
Britannica Online	-->	Wikipedia
personal websites	-->	blogging
evite	-->	upcoming.org and EVDB
domain name speculation	-->	search engine optimization
page views	-->	cost per click
screen scraping	-->	web services
publishing	-->	participation
content management systems	-->	wikis
directories (taxonomy)	-->	tagging ("folksonomy")
stickiness	-->	syndication

Figure 2.4 The key differences between Web 1.0 and Web 2.0 (O'Reilly 2005a)

Gillespie (2008) argues that O'Reilly, 'draws a term from the computational lexicon, loosens it from the specific technical meaning, and layers onto it both a cyber-political sense of liberty and an info-business taste of opportunity'. One would not be surprised therefore to learn that the notion of a move from government to e-government, or perhaps more cynically, e-governance, is also often included in the Web 2.0 platform.

At the beginning of its conceptualization, Web 2.0 was a blanket term for a constellation of often unrelated technologies. It was less of an organised taxonomy and more a list of desirable features, described by O'Reilly (2005a) as 'more of a set of principles and practices'. This was, O'Reilly argued, visualising the web as a platform for making other things. It was a call to arms, a description of the promised land rather than an arrival, and as such it had a firm basis in reality but an end that was very much in the realm of romanticism.

Web 2.0 is less a thing than a brand-name for a disparate collection of ideas and technologies that are gathered together under the term. When O'Reilly, a book publisher, technology evangelist and one of the original 'open source pigs' (Metcalfe 2004), introduced the notion of Web 2.0 he was merely trying to use a common technique in computer software production of 'versioning' the discussion that was taking place. In effect he was posing a question–if the existing constellation of technologies we call the Internet was to be conceptualised as a 1.0 product (i.e. the first version that has a number of bugs and problems that can be ironed out in later versions) then what would the version 2.0 look like. O'Reilly wrote:

> Web 2.0 is the network as platform, spanning all connected devices; Web 2.0 applications [are] delivering software as a continually–updated service that gets better the more people use it, consuming and remixing data from multiple sources, including individual users, while providing their own data and services in a form that allows remixing by others, creating network effects through an 'architecture of participation,' and deliver rich user experiences (O'Reilly, quoted in Schloz 2008).

Of course, as soon as the term was used it was seen as a remarkable way of conceptualising something that had been bugging many of the technologists involved in designing Internet software and services – namely that the present Internet technologies were extremely limited in their dynamism, being essentially static technologies that presented information to the user but which constantly had to undertake slow and repetitive actions to-and-from the server in order to produce the information the user required. The Web 1.0 was built on a model called Client-Server technology which was itself once a cutting edge technology, when technology was understood as being limited to the internal direction of a central organising authority such as a corporation. The Internet, on the other hand, had been built from the bottom-up as a peer-to-peer networking technology – it was designed to be able to allow any user to communicate with any other user – or to use more technical language – each node could communicate with any other node.

This pointed to the fact that the existing Internet technologies, and particularly the web, had been built when web servers were expensive technologies and each client would have a slow connection to the Internet, usually via dial-up modems. The effect of this technical restriction meant that the web server programmers and web site designers had generally internalised the norms of a web restricted by limitations in the older technologies. They had designed based on existing notions of what could be done, rather than with a notion of more powerful technologies and communications being available in the future. This produced a version of the Web that had underlying protocols and plumbing of the Web which had remained within the Client–Server paradigm of exchanging information – HyperText Transfer Protocol (HTTP) used for web pages is a good example of this.

These issues had been bubbling under the surface for quite a while and there had been a number of new technologies developed which had extended the possibilities of the underlying peer-to-peer nature of the lower levels of Internet protocol. Many of these were highlighted by a number of examples that changed the way in which the web was conceptualised: (i) the original version 1.0 Napster, which allowed users to share their music collection with other Napster users. Unfortunately for Napster, and fortunately for the music corporations, Napster still had a model of Client–Server in mind when designing their system which meant that when the corporations successfully bankrupted Napster, they could also shut down the servers that facilitated the network; (ii) equally too, governments who wished to regulate or control the flow of information that passed both into and out of their country, not to mention between users of the Internet, soon realised that to control the web servers was to control the flow of information over the web. It was a simple matter to license the web servers, much in the same way that previously printing presses had been licensed. For the technology evangelists who had foretold that 'information wants to be free' this was a difficult fact to accept; lastly, (iii) for those who designed technologies there was an increasing problem with web 'real estate'. That is, squeezing large quantities of information onto small computer screens was becoming increasingly challenging, taken together with the inefficiency of continually downloading web pages from web servers when data needed to be refreshed (which added to bandwidth and couldn't provide real-time information). There was a recognition that the web was not the dynamic medium that was promised, rather it provided a form of web-based 'Polaroid' picture which soon went out of date.

This discourse of Web 2.0 has spread from the technical sphere into many related areas and in this translation Web 2.0 has become firmly associated with notions of participation, social networking, sharing, open source and free software, open access to knowledge and democratic politics. This bundle of concepts has been extremely powerful and has triggered many groups and individuals to try to build sites that leverage the access to information that the web grants, together with the collaborative spirit of online user groups to sort and process information (referred to as tagging or folksonomies) in order to share information with the public. This in turn has influenced political discourse, which has drawn on this technical background for a language with overtones of progress, modernity, efficiency and high-standards. The lack of a specific definition for the Web 2.0 term has allowed the concept to signify more than it means in itself. As Silver argues,

> There's something quite brilliant, from a corporate–consumer–marketing perspective, about the term Web 2.0. Its very name – Web 2.0 – embodies new–and–improvedness: a new version, a new stage, a new paradigm, a new Web, a new way of living. Attached to any old noun, 2.0 makes the noun new: Library 2.0, Scholarship 2.0, Culture 2.0, Politics 2.0 (Silver 2008).

But what remains interesting about the concept of Web 2.0 is the fact that: (i) its radical break evinced in 2004 bears little or no relation to any discernable empirical evidence of such a break. As many commentators, including Tim Berners-Lee, have pointed out, the technologies that make up the Web 2.0 phenomena predate the announcing of its emergence (see Scholz 2008). (ii) The notion that the technology has such a wide and discernable impact on so many aspects of life and economics draws its force from a rather simplistic notion of technological determinism. (iii) Web 2.0 in essence is another form of announcement for the information society and new ways of structuring aspects of society with networked models of organisation. (iv) Implicit within Web 2.0 is an underlying libertarian ideology that valorises the contribution of the individual (as a rational actor) whether through such notions as collective intelligence, or through the idea of a long-tail economy, or the wisdom of crowds.

Web 2.0, then, has to be understood as a particular constellation of code-based technologies. Although interesting in terms of its focus on increasing interactivity and real-time delivery of data, Web 2.0 does not represent something outside or beyond existing ways of understanding code, indeed, it highlights the importance of critical approaches to new movements and fashions within programming and technology.

Understanding code

Any study of computer code should acknowledge that the performativity of software is in some way linked to its location in a (mainly) capitalist economy. Code costs money and labour to produce, and once it is written requires continual inputs of energy, maintenance and labour to keep functioning. This has important implications when it is understood that much of the code that supports the Internet, even though it is free software or open source, actually runs on private computer systems and networks (see Berry 2008). Should these private corporations decide to refuse to carry public data traffic, or implement prioritising data traffic, this would severely disrupt the peer-to-peer architecture of the Internet. This is currently part of the ongoing debates over the importance of Network Neutrality – whereby private networks agree to carry public and private data in reciprocal arrangements which are implemented in code.[19] It also highlights why the political economy of software cannot be ignored.

The growing importance of intellectual property rights also provide new insights into a cultural politics involving possession and dispossession of a proliferation of digital media – particularly where it facilitates the experience of the user of audio, music, textuality and mass-produced imagery. Code constructs the relationship we have with technology, and it is here where questions of ownership, through patents and copyrights, for example, and technologically mediated control, through digital rights management, become key issues. As I argue elsewhere (Berry 2008),

> [Digital rights management software (DRM)] prevents users from carrying out unauthorised actions on copyrighted works often irrespective of the ownership or rights of the individual user (Lessig 1999). Adobe Acrobat and E-paper, for instance, have the ability to prevent the user from copying, changing and even quoting from a protected document when using the particular DRM-protected software in which the document is delivered to the user. The software is delegated the legal restrictions of the copyrighted work and then prescribes these restrictions back on the user. The user is thus unable to perform activities that break the terms of the legal copyright (Berry 2008: 30–1).

Whilst not wishing to ignore these important, and indeed critical, dimensions of understanding the relationship between software and the wider society, in this book I will largely bracket out the question of political economy of software (I have explored some of these issues in Berry 2008). I do this mainly to tackle the question of the materiality of code in its specificity and to try to think through code as a form that is amenable to a phenomenological encounter but as will be seen, political economy always remains on the periphery of this analysis.

I am exploring code as socio-technical assemblages that are more or less socially embedded in broader networks of social relations and institutional ensembles. Whilst a more phenomenological approach to code is undertaken throughout this book, my aim is to concentrate on the materiality and concreteness of code and highlight the constraints that operate upon it, which may of course involve questions related to the production, distribution and consumption of code as software. Code provides an interesting locus of exploration of the network as an organisation form, a key trope in Information Society discourses which proclaim a new era in the economy and society (Berry 2008: 43–7). At all levels of the network, software and code may be connected to each other in quite counter-intuitive ways; for example, code itself has an internal networked topology, that is, code is not 'above' or 'below' other code, rather code is added to other code as a connection. We sometimes find it easier to understand code through a hierarchical relationship, but strictly speaking code lies on a plane of immanent connections and consequently, no code is 'bigger' or 'more important' than another, except to the extent that it has a larger number of connections. In a political economy of the information society, a more nuanced understanding of the way in which power is located in the network, for example through connections, or through protocol (Galloway 2006), demonstrates that we need to take account of the way in which software as *dispositifs socio-technique* (socio-technical devices) acts to perform the network form (Callon 2007).

Whilst the previous analytical distinctions of code help us to understand and work through the ways in which functional differentiation takes place *within* code and in its development and operation, they are limited in that they are analytical divisions and hence remain on the plane of immanence of code as a medium. We must remember that we should keep in mind that code-based devices are a 'tangle, a multi-linear ensemble' (Deleuze 1992, 159). Code is therefore composed of different sorts of 'lines' – as modalities of execution, internal normative states and positional calculative logic. Code follows directions; it 'traces' processes interacting with its own internal state, the internal state of other code and through external mediating processes, such as the graphical user interface, the user sitting in the real world. But this code 'abstracts' or dis-embeds a logic, whether as here analytically understood as engine, container, image or communications channel.

Code is a complex set of materialities that we need to think carefully about in turn. From the material experience of the user of code, both programmer and consumer, to the reading and writing of code, and then finally to the execution and experience of code as it runs, we need to bring to the fore how code is computational logic located within material devices which I will call technical devices. Now we turn to look at some concrete examples of the materiality of code and how we can understand it as both written, read and executed.

3
Reading and Writing Code

So far I have discussed the difficultly of understanding code and software, and how it is important to link together the materiality of code with its social practices in order to help us understand it. One of the biggest problems with trying to understand code is finding the right kinds of examples to illustrate this discussion so here I will present some examples of code that will make the code more visible and show why reading code is useful. Across the Internet, there are now countless code snippets and repositories containing sample code that demonstrate everything from 'Hello, world!', traditionally the first programming example taught in computer science, to complex database management systems, or even complete operating systems, like Gnu/Linux. Whilst I do want to stress the importance of connecting the dots associated with code and software, I also want to avoid a tedious programming lesson in what can soon become a rather dry subject of discussion. For that reason, I have tried to pick examples that will be either immediately clear as a demonstration of the point I wish to make, or else interesting in their own right. Secondly, I have tried to be clear that when one is discussing code one should be aware that snippets of code can be difficult to understand when taken out of context and often require the surrounding documentation to make sense of it. For example, theoretical study of computational methods often focus solely on the programming logic itself whilst paying no attention to the interesting information that might be found in the commentary sections, documentation, variable names, function names, etc.

In this chapter, we will be looking more closely at the distinction between 'code' and 'software'. Throughout, I shall continue using 'code' (as delegated code) to refer to the textual and social practices of source code writing, testing and distribution. In contrast, 'software' (as prescriptive code) will refer to the object code, that is, code that has been compiled into an executable format, which includes final software products, such as operating systems, applications or fixed products of code such as Photoshop, Word and Excel. In discussing the code and software, we will also be looking at the cultural practices that surround the use of it. This further allows us to think about the processes of software creation through computer programming, and also the related issue of hacking as the clever and sometimes inspired use of programming to change the normal execution of code or of subverting its intended (prescribed code) functions.

In this sense, I will not be focusing on the level of the screen, so-called screen essentialism, rather, the analysis will focus on how the code is constructed by its relationship to a running machinic assemblage. Nor shall I be making literary readings of the textual form of code, as it is crucial that we keep the double articulation of code, as both symbolic and material, fully in view. The way in which the code is created, and tested, maintained and run, forms part of our attention to the materiality and obduracy of code. Thus, to echo the discussion made in the last chapter, we will remain attentive to code in its multiplicity, that is as a literature, a mechanism, a spatial form (organisation), and as a repository of social norms, values, patterns and processes.

Tests of strength

To locate the materiality of code, I develop Latour's (1988) notion of 'trial of strength' introduced in *Irreductions*. Here a test can be considered to be legitimate as long as strengths are being measured according to the tenets of a set of rules. In opposition to the linguistic turn, the notion of a test inclines towards realism. Indeed, for persons to be able to reach an agreement on software in practice, not only in principle, a reality test has to take place, accompanied by a codification or, at least, an explicit formulation of valid proof. Each claim is therefore associated with a series of tests that can be called upon to support its claims. An overriding requirement is the obligation to specify the type of strength that is involved in a specific test and to arrange a testing device. To fail the tests indicates that the 'concrete fact' of the software has failed to be proven and consequently that the software itself remains vapourware, that is unrealised or immaterial. The notion of a test of strength is also similar to the idea of a 'test case' in software engineering, which is a single problematic that can be proved to be successful, and therefore designates the code free from that error or problem. These tests form the basis of the testing the 'realness' of the software, for to fail the tests indicates that the 'fact' or reality of the software has not been achieved. To be included in a particular 'society of code' then, the code must be legitimated (realised) through a series of tests. Code is more visible the more connections it has.

This echoes the work of Gabriel Tarde, a sociologist from the early 20th century, who dreamed of following the actors in a social formation through mapping the connections they made. For Tarde, everything is an association, 'everything is a society' (Latour 2002: 118). This is the starting point of the analysis of software that I want to explore. This will allow us to map the attachments and solidarity that is formed between software and hence trace the way in which it is materialised and made. Boltanski and Thévenot (2006) make a useful distinction between two different test modes: (1) tests of strength (*épreuves de force*), which themselves problematise the boundary conditions and (2) legitimate tests (*épreuves légitimes*), that are tests within the boundary conditions. In a test of strength, it is acceptable to mobilise any and all kinds of strength. Nothing is specified beforehand. Anything goes, as long as it is crowned with success (this could be thought of as hacking code). In a legitimate test, playing by the rules of the situation constrain as to what can be done to be successful (this can be thought of as following legitimate software processes). A legitimate test must always test something that has been defined, presenting itself as a test of something.

To look at the issue of code in more detail and to bring to the fore the element of materiality, I therefore use Latour's notion of the 'trial of strength' to see how the materiality of code, its obduracy and its concreteness are tested within computer programming contests. This approach is useful, as it incorporates both discursive and extra-discursive dimensions in the analysis of code. This is because one of the most important requirements for the materials which make up a programming assemblage is that they pass the predefined 'requirements' tests. These are usually identified in the requirements specification, which is essentially a thin description of the software that is to be created. From this a thick description is then assembled, usually in the design phases, which is a detailed outline of much of the processing that needs to be undertaken, and the general features of the structure, sometimes through pseudo-code or through one of the formal languages such as UML (Unified Modelling Language) or Z. It is only after this point that the prototyping and testing phases really begins and code is written, but it remains an iterative process to construct the detailed structure and content of the required software system.

For example, a series of tests are required to be passed before the software is released to the public. Each stage of the release cycle is materialised in different ways, such as compilation (on hardware), printed (on paper), tested (by humans), or distributed (on a physical medium). As most software continues to be developed in a variant of the 'waterfall model' of software development that runs through a cycle of design-code-test-release software, we can trace this materiality as 'trials of strength'. So software goes through a series of phases which are iterative with the intention of improving software quality with each cycle. Each step creates physical entities (e.g. documentation) and tests that further reinforce the materiality of code. Some of the development cycles that are used include: (i) Alpha, firstly as white-box testing, sometimes called glass box testing, is where the internal logic of the code is tested. When this is complete the testing moves to treating the internals as hidden and black-box testing begins, which seeks to test the external relationships with the code. When the alpha test is complete the code is usually 'frozen' and considered feature complete and no major additions of functionality are added; (ii) Beta, which may be open or closed to the public, and which involves usability testing to ensure that problems for users are minimised. Some companies, especially those associated with web 2.0, use this exclusivity to promote their software and grow the user-base; (iii) Release candidates (RC), where the software has now reached the point at which it is ready for release and undergoes a last phase of tests to iron out any major problems with the software. This version is usually considered 'code complete' and will not undergo any significant additions to the code; and finally (iv) release to manufacturing, otherwise known as release to marketing, where the software is ready to release as a final product. When software is released on CD or DVD this was sometimes known as 'going gold' or 'gold master', a term particularly associated with Apple Corporation. This term is used because the master version delivered to the manufacturers as the final version would be in the form of a gold disk.

Only when a series of tests has then been passed is the final code, now compiled into software, considered a 'release candidate' and therefore can either be given to people as a beta (usually a pre-release version that may still contain semi-major bugs) or else as a final release candidate version close to shipping. Each stage of the process can be understood in terms of the notion of 'trials of strength' allows us to focus on the very real requirements that are made of the code at all stages of its evolution as it is assembled into the final release candidate as released software. In the following sections I want to look at the tests of strength demonstrated in a number of programming examples, for example, in the Microsoft example below, we will be looking at how Microsoft uses the notion of a daily compilation of the entire operating system as a means of testing the

68 *The Philosophy of Software*

fact that the operating system is both functioning and moving forward in development. Throughout these case studies, the intention is to link the symbolic level of the literate programmer with the machinic requirement of compilation and execution of the software.

Reading code

The leaked microsoft source code

In February 2004, Windows 2000 source code was accidentally leaked onto the web, as 'files [dated] 25 July 2000. The source was contained in a Zip file of... 213,748,207 bytes, named windows_2000_source_code.zip, [and] had been widely circulated on P2P networks' (Selznak 2004). This gave anyone on the Internet a rare opportunity to make a close reading of the software written by the programmers at Microsoft. Most of the attention was focussed immediately not on the code, but rather on the colourful commentary that explained how the code worked. This was due to a number of reasons, firstly, the concern over possible copyright infringement of Microsoft's source code, and by not explicitly 'reading' the code programmers hoped to protect themselves by keeping a 'clean room' separation from themselves and the actual code.[1] Secondly, the code runs to millions of lines of very complicated code and the commentary is a form of documenting that is often a first step into the process of understanding it.

Microsoft is renowned for compiling, or 'building', its entire operating system everyday to check that the code works and to root out problems and errors as early as possible in the coding process. The daily build also acts as a disciplinary mechanism on its staff as it puts them and their work under constant observation. As everyone suffers when the build fails there is considerable peer pressure not to 'break the build' by submitting untested or problematic code. In terms of the analysis we are undertaking here, we might think of the build as an important 'test of strength' for the materiality of Microsoft Windows development software. To become part of the Windows source, and therefore to be in the running to be a releasable part of the operating system, requires that you can be materialised in the build through passing the compilation process. This compilation process involves both a material and social aspect, (i) the code must compile on the build machine within the stated constraints of the build process, which are essentially a series of compilation tests; (ii) the entire organisation peer reviews the build process and to cause the build to fail is noticed by the entire community of developers and socially frowned upon; (iii) the daily build usually includes a number of 'smoke tests', or build verification tests, which give 'shallow and wide' means of assessing the new versions of the software. These might include automated scripts to ensure that the correct version of the code was compiled and included, that basic error-checking is working, or that the program does not crash the system, or itself, on launch. It would usually fall to later stages of testing, either under alpha or beta testing for more in-depth testing to be carried out, usually by specialised human testers.

By looking at the code in these files, an insight into Microsoft's daily build process is given. For example, in a message to programmers who might change their code without thinking of the consequences for the build process in the file private\windows\media\avi\verinfo.16\verinfo.h, was

```
 * !!!!!!!!!!!!!!!!!!!!!!!!!!!!!!!!!!!!!!!!!!!!!!!!!!!!!!!!!
 * !!!!!!!!!!!!!!!!!!!!!!!!!!!!!!!!!!!!!!!!!!!!!!!!!!!!!!!!!
 * !!!!!!!!IF YOU CHANGE TABS TO SPACES, YOU WILL BE KILLED!!!!!!!
 * !!!!!!!!!!!!!!!DOING SO F*CKS THE BUILD PROCESS!!!!!!!!!!!!!!!
 * !!!!!!!!!!!!!!!!!!!!!!!!!!!!!!!!!!!!!!!!!!!!!!!!!!!!!!!!!
```

Figure 3.1 Microsoft Windows source code commentary (Selznak 2004)

Microsoft programmers also revealed their feeling about several 'moronic' moments in the code where they point to errors or bugs introduced by other programmers (we could think of these as 'soft' tests of strength). This points not only to 'moronic' programming practices, but to the problems any organisation will have in managing projects with many tens of millions of source lines of code. Not only is there a problem in holding the project in any individual's head, but the added complication of staff leaving, corporate memory therefore being drained, and, additionally, new software being added to old software which was never designed to be extensible in quite the way that would have been helpful to later developers. Of course, there will also be a dimension of blaming other departments or software team for their 'moronic' decisions in design and implementation. This example demonstrates the necessity of communication and good high-level system architecture planning in any large software project:

```
private\genx\shell\inc\prsht.w:
// we are such morons. Wiz97 underwent a redesign between IE4 and IE5
private\shell\ext\ftp\ftpdrop.cpp:
We have to do this only because Exchange is a moron.

private\shell\shdoc401\unicpp\desktop.cpp:
// We are morons. We changed the IDeskTray interface between IE4

private\shell\browseui\itbar.cpp:
// should be fixed in the apps themselves. Morons!
```

Figure 3.2 Microsoft Windows source code 'moron' comments (Selznak 2004)

The documents also showed where Microsoft employees were required to break programming conventions and perform 'hacks', or inelegant software fixes to get around stupid, restrictive or problematic bottlenecks in the existing codebase. Professional programmers, it goes without saying, should not be resorting to hacks to get code to work as they inevitably generate more problems and require more hacks to fix in the long term. Hacks are different to 'moronic' moments in that they are temporary fixes to make things work – they are often intended to be removed at a later date. Hacks are also often implemented in response to urgent need, perhaps a major problem has been identified in the code very close to the release date, or a security flaw has been uncovered. This is a very bad way of dealing with code issues as it runs the risk of importing even more bugs into code than those it was meant to deal with, however, sometimes there is just no way of avoiding the need for an urgent patch for code, even if contrary to expectations it then becomes a permanent repair that is soon forgotten about.

```
rivate\ntos\w32\ntuser\client\dlgmgr.c:
// HACK OF DEATH:


private\shell\lib\util.cpp:
// TERRIBLE HORRIBLE NO GOOD VERY BAD HACK


private\ntos\w32\ntuser\client\nt6\user.h:
* The magnitude of this hack compares favorably with that of
the national debt.
```

Figure 3.3 Microsoft Windows source code 'hack' comments (Selznak 2004)

By reading the Microsoft source code one also begins to get connections to the political economy of software development more generally. For example, Microsoft uses certain specialised function calls called Application Programming Interfaces (APIs) which are kept private and internal to the company. These are then used by its own software which, it is alleged, give it a performance boost over its rivals third-party software. Whilst not technically illegal, they certainly demonstrate the advantages of monopoly control of a software platform that can be turned to profitable advantage, for example,

```
private\mvdm\wow32\wcnt132.c:
// These undocumented messages are used by Excel 5.0

private\windows\shell\accesory\hypertrm\emu\minitelf.c:
// Ah, the life of the undocumented. The documentation says
// that this guys does not validate, colors, act as a delimiter
// and fills with spaces. Wrong. It does validate the color.
// As such its a delimiter. If...
```

Figure 3.4 Microsoft Windows source code 'undocumented' comments (Selznak 2004)

Undocumented features can also be used to short-cut programming by allowing tricks to be used in software (as shown in the second example above). Again, these are risky manoeuvres as they may be easily changed in future, and being undocumented by their very nature means that the owners of the undocumented effect may not realise that others might be using it. These examples give a clear picture of the rather murky world of programming, particularly on large-scale and legacy systems where the actuality of programming day-today may involve more decoding, hacking and use of undocumented features than one might expect from what on the outside look like professional programming practices.

Much of the media attention on the leaking of the source code focussed rather oddly on how the software might be stolen or used by rivals, for example the BBC reported that 'such access could provide a competitive edge to its rivals, who would gain a much better understanding of the inner workings of Microsoft's technology' (BBC 2004). This is unlikely, as not only was the source code incomplete, it was also largely out of date, and software is continually and dynamically under development with the code tree in a constant state of flux. Betting your company on the use of private APIs or esoteric or undocumented functions would be to trust your company to fate and hope that Microsoft didn't change something at a later date. Programmers themselves had a field day searching for rude words, general items of interesting code, and checking for whether rumours of bad practices and poor programming were true (see Slashdot 2004). Many of the commentators remarked on the unlikely situation of anyone finding much of interest to take from Microsoft code, pointing to the difference between having the source code as a textual repository and actually getting it to compile. When it is remembered the code runs to millions of lines, and the compilation build process is doubtless idiosyncratic and specific to Microsoft itself, it is clear that the codebase would challenge the memory of any single programmer to unpick. CNN reported that the 'leaked Windows 2000 code contained 30,915 files and a whopping 13.5 million lines of code' (Legon 2004), other experts have calculated this to be approximately 47 per cent of the total codebase of 29 million lines of code (Jones 2004).

The leaking of the code also led to amusing parodies of the source code such as the fragment below,

```
if (detect_cache())
    disable_cache();

if (fast_cpu())
{
    set_wait_states(lots);
    set_mouse(speed, very_slow);
    set_mouse(action, jumpy);
    set_mouse(reaction, sometimes);
}
/* printf("Welcome to Windows 3.1");     */
/* printf("Welcome to Windows 3.11");    */
/* printf("Welcome to Windows 95");      */
/* printf("Welcome to Windows NT 3.0");  */
/* printf("Welcome to Windows 98");      */
/* printf("Welcome to Windows NT 4.0");  */
printf("Welcome to Windows 2000");

if (system_ok())
    crash(to_dos_prompt)
else
    system_memory = open("a:\swp0001.swp", O_CREATE);
```

Figure 3.5 Parody of the Microsoft Windows source code (Baltimoremd n.d)

This fragment refers to the fact that users and programmers have long complained about the slow operation of the Microsoft operating system, so for example the parody disables the cache, which would slow down the computer, and sets the mouse speed to 'very_slow'. The 'code' also sarcastically points to the genealogy of Windows in its different versions (e.g. 3.1, 3.11, 95, etc.) and insinuates that every new version of Windows is really the same operating system renamed (e.g. the printf commands).[2]

Climate research code

Another example of the way in which open sourced code and crowd-sourcing on the Internet can come together in fascinating ways around source code concerns the appearance on the Internet of over 1,000 private e-mails containing data and code from the University of East

Anglia concerning climate data. This became a matter of concern, known as 'climategate', a controversial data series that was distributed when emails were stolen from the University of East Anglia Climatic Research Unit (CRU). In particular, this grew into what became known as the 'Hockey Stick' controversy. The *Guardian* explains '[t]he "hockey stick" graph shows the average global temperature over the past 1,000 years. For the first 900 years there is little variation, like the shaft of an ice-hockey stick. Then, in the 20th century, comes a sharp rise like the stick's blade' (Pearce 2010). The data and emails were immediately shared, commented on, and subject to a great deal of debate and controversy due to the way in which the researchers appeared to be cavalier with the data. As the CRU produces one of the four most widely used records of global temperature and these have been key to the Intergovernmental Panel on Climate Change's (IPCC) conclusions that the planet's surface is warming and that humanity's greenhouse gas emissions are very likely to be responsible, it is easy to understand why a controversy could soon erupt over perceived bias in the scientific method.

> Although it was intended as an icon of global warming, the hockey stick has become something else – a symbol of the conflict between mainstream climate scientists and their critics. The contrarians have made it the focus of their attacks for a decade, hoping that by demolishing the hockey stick graph they can destroy the credibility of climate scientists (Pearce 2010).

Eric Raymond, a key activist in open source software and a critic of theories of climate change, demonstrated this by showing the data and the code that processes and applies the data series in the FORTRAN code stolen from the CRU on his blog and commenting upon it,

> From the CRU code file osborn-tree6/briffa_sep98_d.pro, used to prepare a graph purported to be of Northern Hemisphere temperatures and reconstructions.

```
;
; Apply a VERY ARTIFICAL correction for decline!!
;
yrloc=[1400,findgen(19)*5.+1904]
valadj=[0.,0.,0.,0.,0.,-0.1,-0.25,-0.3,0.,-
0.1,0.3,0.8,1.2,1.7,2.5,2.6,2.6,$
2.6,2.6,2.6]*0.75 ; fudge factor
if n_elements(yrloc) ne n_elements(valadj) then message,'Oooops!';
yearlyadj=interpol(valadj,yrloc,timey)
```

This, people, is blatant data-cooking, with no pretense [sic] otherwise. It flattens a period of warm temperatures in the 1940s 1930s — see those negative coefficients? Then, later on, it applies a positive multiplier so you get a nice dramatic hockey stick at the end of the century (Raymond 2009).

Whilst not a comment on the accuracy or validity of Raymond's claims, this example shows the importance of seeing the code and being able to follow the logic through code fragments that can be shared with other readers. A large number of comments soon attached to this post, critiquing the code, checking the data and seeking clarification about how the CRU had used this data. This controversy wasn't only localised here, however, the code critiques spread across the Internet and discussion forums with many eyes casting a critical gaze over fragments of the source code. Especially files such as 'HARRY_READ_ME.txt' (Harry 2009), which contained a commentary by a programmer on the project including his thoughts and mistakes, which many took to be a smoking gun (KinsmanThoughts 2009). Although it later turned out that this file was actually rather unremarkable.

Crucially though, and this is a running theme, the code could also be downloaded, compiled and run, and the actual processing of data analysed to see if outputs like these are producing accidental or deliberate artefacts that were distorting the data. Not only is this a clear example of the changing nature of science as a public activity, but also demonstrates how the democratisation of programming means that a large number of people are able to read and critique the code.

After a detailed seven month investigation, the Independent Climate Change Email Review led by Sir Muir Russell, a former civil servant and former vice-chancellor of the University of Glasgow, it was found that the rigour and honesty of the CRU researchers as scientists were not in doubt (Black 2010). They did, however, comment on the way in which science must increasingly defend itself in the public arena and make available its data and code underlying its conclusions to avoid this kind of controversy in the future. The committee pointed to the new way in which the Internet raises 'important issues about how to do science in such an argumentative area and under new levels of scrutiny, especially from a largely hostile and sometimes expert blogosphere' (*The Economist* 2010b). Nonetheless, the way in which the controversy played out across the Internet, with many different actors checking the code, critiquing it and trying to discover the accuracy of the projections, showed how important code is becoming for our understanding of the world, and for political policy responses to it.

Writing code

To look at the questions raised by writing code, I now focus on two critical case studies in the remainder of this chapter, whose processes allow us to see the generalised way in which software development is a continual reflexive activity. These cases also allow us to see how the materiality of code is demonstrated by abiding closely to the prescribed legitimate tests for the code being developed. In some cases allowances are made for considerable creativity and bending of the rules in order to achieve the system design that is required. In other cases, the materiality is shown precisely where the tests of strength can be implemented which completely undermine or bypass these legitimate tests, and which can change the whole nature of the testing and development process.

The first case study is the Underhanded C Contest, an online contest that asks the contestants to submit code that disguises within fairly mundane source code another hidden purpose. The second case study is The International Obfuscated C Code Contest (IOCCC), a contest to write the most Obscure/Obfuscated C program possible that is as difficult to understand and follow (through the source code) as possible. By following the rules of the contest, and by pitting each program, which must be made available to compile and execute by the judges (as well as the other competitors and the wider public by open sourcing the code), the code is then shown to be material providing it passes these tests of strength. Each of these competitions raise a series of requirements and tests that code has to meet in order to fulfil the requirement of being described as being code at all.

The Underhanded C Contest

The Underhanded C Contest is an online programming contest in which contestants aim to create code that gives the appearance of performing one function, whilst subtly and preferably invisibly doing another. The software must be of a high standard in terms of literate programming, that is, 'the main goal… is to write source code that easily passes visual inspection by other programmers' (XcottCraver 2008). But as the competition website explains,

> In this contest you must write code that is as readable, clear, innocent and straightforward as possible, and yet it must fail to perform at its apparent function. To be more specific, it should do something subtly evil (XcottCraver 2008).

The competition has been running since 2005 and is organised by Dr Scott Craver of the Department of Electrical Engineering at Binghamton University. The code is required to be in the C programming language and must be compilable and supplied in a form that can be demonstrated both textually, as the source code, and mechanically, as an executable process on the machine following compilation. The key to the underhand competition is that the underhanded behaviour has to be in the code itself. There is therefore a strong element of human deception required in these programs, because rather than 'hacking' a protocol, you must mislead a programmer. The misleading, can take many forms, but one of the best is to ensure that nothing that is done by the programmer is out of the ordinary. This relies on the fact that the programmer/tester will expect a certain kind of predictable behaviour from the programming code and this can be exploited by the underhand programmer. For example, if the output that results from the execution of the program doesn't meet the requirements of what would expected as 'standard output' it would probably be regarded as suspicious. The goal is to engineer malicious behaviour that is not noticed as part of a test or code review.

Here I will focus on the fourth annual contest (12 June 2008–20 September 2008). The reason for discussing the 2008 entry was that in previous years the programming requirements tended to be rather complex and esoteric, whereas in 2008 the requirement was merely to black out, or redact, a standard image file on the computer. The simplicity of the requirements disguised a rather difficult task, namely to redact a file whilst allowing the redaction to be undone, whilst keeping the code relatively clear for peer review. The contest used a digital data structure called 'PPM file in ASCII (P3)' format, this meant that the saved format of the image file was in a textual format that made it simpler for programming calculation.

The contest had over 100 entries by programmers and the formal requirement was that entries,

> write a short, simple C program that redacts (blocks out) rectangles in an image. The user feeds the program a PPM image and some rectangles, and the output should have those rectangles blocked out (XcottCraver 2008).

The idea is that the redacted content from the image file is somehow not actually wiped. Ideally, the image would appear blocked-out, but somehow the redacted blocks could be resurrected. This is the principle of leaky redaction, so that with a small amount of effort the leaked data should be able to be reconstructed. As part of the formal requirements, the command line code that would be needed to execute it was also specified to ensure that the clever part of the programming would take place within the C program that was submitted, rather than through mysterious batch files or another mechanism.

```
gcc -o redactomatic obviouslyinnocentprogram.c
% redactomatic in.ppm > out.ppm
10 14 121 44
10 60 121 90
10 104 121 134
^D
```

Figure 3.6 Redacting command line execution

The rules of the redaction code design were simple: (1) the program should be compiled easily; (2) the user can then type in values that correspond to rectangles that should be redacted as x,y coordinates; (3) the user exits (^D) and the file is output (out.ppm). As can be seen these are the boundary conditions for the materiality of the code to be demonstrated, as part of the testing of the code the judges would compile the code themselves on a suitable machine, and they would run the test on a number of image files.

The legitimate tests, included the key objective of the competition: to produce a source file that looks very innocent, and passes informal code inspection. The additional legitimate tests that are used to validate the software were: (i) 'Short programs are innocent, and more impressive. If your source file is over 200 lines, you are not likely to win. You can hide a semi truck in 300 lines of C. In general, the fewer hiding places, the more impressed we will be if you can conceal malicious behavior' (XcottCraver 2008); (ii) 'Typical behavior is innocent. Unusual and unnecessary steps will raise eyebrows unless you can find a reasonable excuse for them. This makes this challenge somewhat difficult, because there are only so many ways a dude can wipe out a rectangle' (ibid. 2008). There was also the potential for extra points: (iii) extra points were awarded if the error, once found, looks like an innocent bug rather than deliberate miscoding; (iv) extra points were awarded if the code still appears innocent under syntax coloring; (v) extra points were awarded if the information leakage is dramatic.

First place: John Meacham

The winning entry was by John Meacham, a programmer from California, in the US. He submitted a 55-line coded entry that took the image file, scanned over the numbers, and zeroed out the redacted pixels in the most obvious way possible. The code replaced numbers with zeroes on the character level. The leakage was due to the fact that low-intensity pixels were replaced with a '0', and high-intensity pixels replaced with a '00' or a '000'. This was completely invisible in the image itself.

```
for(c = buf;*c;c++) {
        if(isdigit(*c)) {
                if(!ws) {    // new number, increment location.
                        ws = 1; x++;
                        if(x >= width * 3) {
                                y++; x = 0;
                        }
                }
                if(x > rx * 3 && x <= (rx + rwidth) * 3 && y > ry && y
< ry + rheight)
                        putchar('0');
                else
                        putchar(*c);
        } else {
                ws = 0;
                putchar(*c);
        }
}
```

Figure 3.7 Underhanded C Contest, winning entry by John Meacham

The way in which the underhand function was created involved cleverly utilising the format of the image file. This format was a plain text format, it had some basic header information, and a list of whitespace separated numbers, such as '234 2 0', where the numbers represented the magnitude of the red, green, and blue component for each pixel to be displayed on the screen.

The redactor program appeared to simply replace values within the target rectangle with zero. However, due to the way the code processed the file character by character, it is able to leak how many digits each value had to begin with. So for example '234 2 0' would be redacted to '000 0 0'. This was completely invisible when viewing the PPM file, all the values count as zero as far as the format was concerned, but by looking at the original file, you could recover some information about what was in the blanked out area. Whilst this saved a certain proportion of any image file, it should be clear that it is particularly effective with black and white documents, the format of most textual image files. That is because only the values of '0' and '255' are used in these black and white files and when redacted it is easy to read the '0' and '000' as the correct values. One can also easily write a small piece of code to convert the '000' back into the correct '255'.

0	255	0
255	255	255
0	255	0

0	000	0
000	000	000
0	000	0

Figure 3.8 Underhanded C Contest, contents are wiped keeping 255 as '000' length, showing how the basic image information is retained after redaction

As part of his explanation of the code and to legitimate the peer-review aspect of the entry, Meacham also supplied a dramatisation of passing the all important code inspection,

Spook: "So why did you process the file character by character, rather than doing the more obvious scanf("%i %i %i",&r,&g,&b) to read in the values?"

Me: "Well, in order to do that I'd have to read in entire lines of the file. Now there is the gets function in C which does that, but has a well known buffer overflow bug if the line length exceeds your buffer size, so I naturally used the safe fgets variant of the function. Of course, with fgets, you can just assume your buffer size is greater than the maximum line length, but that introduces a subtle bug if it isn't, you may end up splitting a number across two buffers, so scanf will read something like 234 as the two numbers 23 and 4 if it is split after the second character, hence the need to consider each character independently."

Spook: "Ah, of course. good job at spotting that."

Me: *snicker*

Second place: Avinash Baliga

The second placed entry uses a different method of achieving a similar result, this is technically interesting because of the use of a buffer overrun. A buffer overrun, or buffer overflow, is when a programmer accidentally writes past the end of a file or variable into adjacent memory on the computer. In standard production systems, this kind of error can cause major problems and is actually a key way of creating 'exploits' to break a secure system (it was the technique used to jailbreak the iPod and iPhone, for example). The designed 'bug' in this submission is in the `ExpectTrue` macro (located at the start of the code) which prints into a small buffer (small because it is redefined in `main()`) overwriting the mask used to zero out the data with a '0×0a'. This method of writing allows two bits to survive the redaction, low enough in intensity to pass visual inspection, but high enough to reconstruct the redacted data later.

```
/* Error checking macro. */
#define ExpectTrue(cond__, msg__)
_snprintf(buf, 255, "%d:    %s", __LINE__, msg__);
if (!(cond__)) {
fputs(buf, stderr);
exit(-1);
}
...
int main(int argc, char** argv)
{
        pixel p = {{0,0,0}};
        int left = 0, top = 0, right = 0, bottom = 0;
        int mask = 0, x = 0, y = 0, z = 0;
        char buf[32] = {0};
        ...
        ExpectTrue( copy_ppm(in, &out),
            "Error: could not allocate output image.n");
```

Figure 3.9 Underhanded C Contest, second place entry by Avinash Baliga

The judges were clearly taken by this submission and awarded this entry,

> extra points for sheer spite, concealing the evil behavior in an error checking macro. Spite will always get you extra points in the Underhanded C Contest. In the final analysis, this guy gets points for style and technical expertise; the only problem is that masking out pixels, rather than zeroing them, is an operation that is difficult to justify (XcottCraver 2008).

This is interesting because the programmer used an error-checking fragment of code to actually implement the underhanded function. This naturally appeals to the idea that the best place to hide such a function is in the most blatant location in the code, where most people would expect the highest level of probity and care to be taken.

Third place: Linus Akesson

The last entry discussed here came third in the competition and was submitted by Linus Akesson. He employed an 'important Underhanded coding principle: make the common case evil, and the uncommon case wrong' (XcottCraver 2008). This programming example relies on the fact that although virtually all PPM files use 8-bit RGB (red/green/blue) values, that is values between 0 and 255, higher values are possible, namely 16 bit values from 0 to 65535. This allows the programmer to pretend he is checking for these 16 bit values, whereas in actuality he is using this check to implement the underhanded function.

```
#define BYTESPERPIXEL(fits8bits) (3 << (!fits8bits))
...
int main(int argc, char **argv) {

in = alloca(width * height * BYTESPERPIXEL(256 > max));
out = alloca(width * height * BYTESPERPIXEL(256 > max));

fread(in, BYTESPERPIXEL(256 > max), width * height, stdin);

ptr = out;
for(y = 0; y < height; y++) {
    for(x = 0; x < width; x++) {
        for(i = 0; i < BYTESPERPIXEL(256 > max); i++) {

            *ptr++ = *in++ & visibility_mask(x, y, argc, argv);
        }
    }
}
printf("P6n%d %dn%dn", width, height, max);
fwrite(out, BYTESPERPIXEL(max < 256), width * height, stdout);
```

Figure 3.10 Underhanded C Contest, third place entry by Linus Akesson

This is perhaps the most complicated submission as it relies on a bug in the BYTESPERPIXEL macro, due to a lack of a pair of parentheses. This means that BYTESPERPIXEL (256>max) is always worth '3', and BYTESPERPIXEL (max<256) is always '6'. Essentially, the images are allocated, read and processed with 3 bytes per pixel and then the output is written with 6 bytes per pixel. The program reads into buffers created on the stack with alloca(), so the in buffer is right after the out buffer, and swapping '256>max' with 'max<256' at the end ensures that both buffers are written to the output file. In this code, the macro BYTESPERPIXEL gives the false impression that the code intelligently supports higher bit widths than we are ever likely to experience. A small side effect of this trick, however, is that trying to redact those larger bit-depth images cause the program to fail completely. Nonetheless, that apparent support for larger images helps to disguise the fact that the 8-bit case is able to leak information into the file.

Now we turn from the hiding of a new function in the code to the desire to render source-code as unreadable as possible. This is a technique called obfuscation and demonstrates both the mutability of source code itself, and the fact that unreadable source code can still be executable.

The International Obfuscated C Code Contest

The International Obfuscated C Code Contest (IOCCC) is a competition for programmers to write the most complicated looking C program in computer code possible. This is code that is as difficult for a reader to understand and follow through the *textual source* as possible. In other words, the intention is to write *illiterate* code, rather than the clear readable code argued for in Knuth's notion of literate programming discussed above. However, it is not the aim to write functionally complicated programs, rather that the programs submitted should be as simple in their functionality as possibility but as difficult to read as can be managed. The IOCCC has been running since 1984 (with a few exceptions), and it was started by Landon Curt Noll and Larry Bassel on 23 March 1984, whilst they were employed at National Semiconductor's Genix porting group, they write,

> we were both in our offices trying to fix some very broken code. Larry had been trying to fix a bug in the classic Bourne shell (C code #defined to death to sort of look like Algol) and I had been working on the finger program from early BSD (a bug ridden finger implementation to be sure). We happened to both wander (at the same time) out to the hallway in Building 7C to clear our heads... We began to compare notes: "You won't believe the code I am trying to fix". And: "Well you cannot imagine the brain damage level of the code I'm trying to fix". As well as: "It more than bad code, the author really had to try to make it this bad!"... After a few minutes we wandered back into my office where I posted a flame to net.lang.c inviting people to try and out obfuscate the UN*X source code we had just been working on (Noll *et al.* 2009).

The rules of the contest set the general outlines of the test of strength for the competition, and demonstrate the code's actuality in terms of the contest. Each of the code submissions is then compared against each other, but they must be able to be compiled and executed by the judges (as well as the other competitors and the wider public by open sourcing the code). The code is then shown to succeed providing it passes these tests of strength. The competition organisers include a handy dictionary definition of 'obfuscate' on the website to guide the programmers as they craft their submissions, which reads:

> **Obfuscate**: tr.v. -cated, -cating, -cates. 1. a. To render obscure. b. To darken. 2. To confuse: his emotions obfuscated his judgment. [Lat.

obfuscare, to darken : ob(intensive) + Lat. *fuscare*, to darken < *fuscus*, dark.] -obfuscation n. obfuscatory adj (Noll *et al.* 2009).

Code obfuscation means applying a set of textual and formatting changes to a program, preserving its functionality but making it more difficult to reverse-engineer. Generally, obfuscated code is source code that has been made very difficult to read and understand. 'Obfuscators' achieve this by altering the textual and functional structure that makes a program human-readable. They also use macro pre-processors to mask the standard syntax and grammar from the main body of code. Obfuscations may also create artistic effects through keyword substitutions or the use, or non-use, of white space, to create patterns or even complete images from the textual code. Obfuscated C Codes are highly creative examples of coding employing the 'C' programming language.

These programs combine an executable function with an aesthetic quality of the source code. Since 1984 there have been programming contests such as the "International Obfuscated C Code Contest" in which the best programmers worldwide compete. The challenge is to employ programming languages like C, C++ and Perl under particularly restrictive rules but in an extremely creative way. The obfuscated C code contest rules are quite simple - "Write, in 512 bytes or less, the worst complete C program". The aims of the contest are to present the most obscure and obfuscated C program, to demonstrate the importance of ironic programming style, to give prominence to compilers with unusual code and to illustrate the subtleties of the C language (Digitalcraft.org 2006).

To demonstrate how a program is obfuscated, here is a simple six line version of 'hello world' code' written in source code:[3]

```
#include "stdio.h"
int main() {
        printf("Hello World!");
        return 0;
}
```

Figure 3.11 Simple example of a C program

This is one of the simplest programming tasks that is asked of programmers new to a language as it shows how to get a simple textual output to the screen, but also the processes of compilation and the look and feel of the language. The simplest way to obfuscate a program is to convert the text string 'Hello World!' into characters held in a special data form called an array. This just makes it hard to read, as you can see in the identical code below, but slightly obfuscated to hide the text,

```
#include "stdio.h"
void myFunction(int array[], int arraySize) {
        int i;
        for (i=0; i<arraySize; ++i) {
              printf("%c", array[i]);
        }
}
int main() {
        int array[]={72,101,108,108,111,32,87,111,114,108,100,33};
        myFunction(array,12);
        return 0;
}
```

Figure 3.12 C program with obfuscated characters with function call

Here, the 'Hello, World!' has been translated into the separate ascii characters that contain a number to reference the letter. So, for example, 'Hello' is rendered as a series of numbers: 72 (H), 101 (e), 108 (l), 108 (l), 111 (o). From this it is clear that obfuscation is the replacement of textual items that are identical for the computer, or are computable to the same, but which to the human eye are difficult to read or follow. Indeed, one of the most popular tricks is to confuse the eye by making the structure appear fragmented or disconnected.

There a number of additional changes that can now be made to make it more difficult to read the text including: (i) making the function recursive so that it calls itself, a notoriously difficult way of thinking in everyday experience; (iii) using indexes and special conditions in the array, such as hiding certain character values; (iii) using hexadecimal or octal numbering instead of decimal, such as '0x64' or '0x6F'; (iv) renaming variables into difficult to read names, such as changing 'integer' to '__' or 'arraySize' to '_0'; (v) renaming function names to be difficult to read, so that 'myFunction' becomes '_'; and finally (vi) changing the formatting by deleting whitespace that helps our eyes follow the text and instead breaks the logic and continuity of the code. In doing this, the aesthetic dimension of the code is also bought to the fore, together with the intricacies of programming style and syntax. It requires an ability to not only craft a suitable program to perform a function, but to think about presentation and visual impact. In this case, for example, this would have the end result of this series of obfuscations:

```
#include "stdio.h"
_(__,_0,O_,___){(O_<_0)?printf("%c",(O_==2)
||(O_==3)||(O_==9)?___++,O_++,108:*((int*)__
+O_++-___)),_(__,_0, O_, ___):0;}main(){int array
[]={72,'e',111,040,0127,0x6F,'r',0x64,041 };_(array
,
12
,
0
,
0
)
;
}
```

Figure 3.13 C program now obfuscated through text changes and confusing formatting

This example, though, is a very basic attempt and would certainly not impress the contest judges. They are looking for real flair and creativity in the use of the above techniques, together with the kind of detailed technical knowledge that demonstrates real programming skill. So, for example, they detail in the overview of the competition the requirement: (i) 'to write the most Obscure/Obfuscated C program'; (ii) 'To show the importance of programming style, in an ironic way'; (iii) 'To stress C compilers with unusual code'; (iv) 'To illustrate some of the subtleties of the C language'; (v) 'To provide a safe forum for poor C code. :-)' (Broukhis *et al.* 2009). The technical rules of the contest are stated as:

1) The entry must be a complete program.
2) The size of your program source must be <= 4096 bytes.
...
7) The program must be of original work.
...
10) Entries requiring human interaction to be built are not permitted.
...
12) Legal abuse of the rules is somewhat encouraged.
13) Your source may not contain unescaped octets with the high bit set, i.e., your source may not contain octet values between 128 and 255 (Broukhis *et al.* 2009).

This set of rules for the competition reinforce the argument that the contest is not about what the program does, rather, it is about what the source-code looks like, it is the reading of the code that will be used as the judge of the best code entry, but nonetheless its materiality is demonstrated by the requirement to be compilable and executable without 'human interaction'. Preferably, it should be as difficult to read as possible, leaving no clue to the reader as to the way in which the program is logically and functionally constructed. However, it should still conform to both the tests of strength and the legitimate tests which are the foundation for the materiality of the code, in other words it must not break any rules. They do however stipulate the literalness of the reading of the rules, and encourage the 'legal abuse' of the rules, so that although test of strength must be abided by, such as the requirement that the code will compile and run, the legitimate tests are open to interpretation, and clever circumvention.

A running thread through the rules and regulations of the contest, however, remains that the code should execute as a 'normal' program in whatever it is that the program does. Additionally the code is made available on the web and presented so that any user can download, compile and run the program, submitting it to a form of peer-review. In other words, the code always remains 'open sourced' so that it can be tested. This visibility of the code, and full documentation about what it does, how to compile, what version of GCC etc are all included, together with the judges decision and reasoning. We'll now turn to look at some examples of previous winners or notable entries to the competition.

Obfuscated code examples

Each of the following examples shows some of the best entries for the obfuscated code competition. They have been selected both for their technical ability and the way in which they have succeeded in meeting the tests of strength outlined in the rules of the contest. However, most interestingly perhaps, is the extra layer of semiotic meaning that many of the programmers choose to add to the source code in terms of visual images embedded within the code – often as a recursive joke or aside. I think this is interesting both in terms of the way in which close reading of the code becomes increasingly difficult, and consequently a distant reading of the code becomes an appreciation of the visual imagery. It is also notable that the judges apply the tests of strength in terms of the entire entry, often remarking on the hermeneutic fit between the visual image and the underlying source code text and function. Banks, whose impressive implementation of a wire-frame playable flight simulator, shown below, is matched by a clever visual representation in the source code, gives a striking example of this.

Entry for 2004 by kopczynski[4]

This example is an implementation of optical character recognition that detects the characters 9, 8, 10 and 11. Whilst its sparseness is remarkable considering what it is designed to do, the shortness of the code shows the programming prowess, but also does so in a single programming line.

Source code

```
main(O){int I,Q,l=O;if(I=1*4){l=6;if(l>5)l+=Q-8?1-(Q=getchar()-
2)%2:1;if(Q*=2)O+="has dirtiest IF"[(I/-Q&12)-
1/Q%4];}printf("%d\n",8+O%4);}
```

Figure 3.14 Performs OCR of numbers 8, 9, 10 and 11

Build Instructions

To build: make kopczynski

Programmer comments

The program proves it is not as hard to recognize numbers. The current one line version should correctly recognize all numbers from 8 to 11. (Unless you give it some very hard cases. For example, it cannot recognize negatives, zeros which have dots or slashes inside them, digits should be separated, and numbers should be "complete".)

Judge comments

What is in a line? A lot when you obfuscate the way Eryk Kopczynski did it. This small one line program is an outstanding technical work of art as well as one of the better one-liner programs that we have seen in years!

Compile and run without arguments. As input, give it an ASCII graphics figure 8, 9, 10, or 11 made of pound signs and spaces, of any size, shape, or orientation (that's right, an upside down 9 is still 9 :-).

Entry for 1984 by laman (prints spiralling numbers, laid out in columns)[5]

This example prints out spiralling numbers to the screen in columns. The 'recursive' joke that the code represents in a visual form the action that it implements is another demonstration of the hacker sense of humour.

Source code

```
a[900];         b;c;d=1          ;e=1;f;         g;h;O;           main(k,
l)char*         *1;(g=           atoi(*          ++l);            for(k=
O;k*k<          g;b=k            ++>>1)          ;for(h=          O;h*h<=
g;++h);         --h;c=(          (h+=g>h         *(h+1))          -1)>>1;
while(d         <=g){           ++O;for         (f=O;f<          O&&d<=g
;++f)a[         b<<5|c]         =d++,b+=        e;for(           f=O;f<O
&&d<=g;         ++f)a[b          <<5|c]=         d++,c+=          e;e= -e
;}for(c         =O;c<h;         ++c){           for(b=O          ;b<k;++
b){if(b         <k/2)a[          c]^=a[          ^=a[(k           -(b+1))
<<5|c]^=        a[b<<5           |c]^=a[         (k-(b+1          ))<<5|c]
;printf(        a[b<<5|c        ]?"%-4d"                         ,a[b<<5
|c]);}          putchar(         '\n');}}        /*Mike           Laman*/
```

Figure 3.15 Prints spiralling numbers, laid out in columns

Programmer comments

NOTE: Some new compilers dislike lines 6 and 10 of the source, so we changed them... I hope you have the C beautifier! The program accepts ONE positive argument. Seeing is believing, so try things like:

    laman 4
    laman 9
    laman 16

This code should run you in circles.

Entry for 2004 by arachnid[6]

The example by arachnid takes as an input ASCII files that are formatted into mazes, that it then allows the user to navigate. Again, a recursive 'joke' is that the source code can be fed as input to the program to play in the same way.

Source code

```
#include <ncurses.h>/***********************************************/
           int           m[256              ] [         256    ],a
 ,b    ;;;    ;;;   WINDOW*w;  char*l=""    "\176qxl"   "q"     "q"    "k"   "w\
xm"   "x"    "t"          "j"         "v"          "u"         "n"          ,Q[
 ]=   "Z"   "pt!ftd`"   "qdc!`eu"   "dq!$c!nnwf"/**   ***   */"t\040\t";c(
int    u       int             v){                       v?m       [u]      [v-
 1]    ]=2,m[u][v-1] &   48?W][v-1   ] &   15]]):0:0;u?m[u   -1][v]|=1  ,m[
     u-       1][       v]&        48?            W-1       ][v         ]&
 15]   ]):0:0;v<   255     ?m[   u][v+1]|=8,m[u][v+1]&   48?   W][    v+1]&15]]
 ):0;         :0;        u         <         255       ?m[    u+1        [v
4,m[u+1][    v]&48?W+1][v]&15]]):0:0;W][   v]&   15]    ]);}cu(char*q){   return
     ?d        ?cu        (q+         1)&         1?q        [0]         ++;
q[0 ]--     :1;     }d(    int     u ,    int/**/v,     int/**/x,    int    int
Y=y    -v,    X=x    -u;    int    S,s    ;Y<    0?Y    =-Y    ,s,
s=-    1:(    s=1);X<0?X=-X,S    =-1   :(S=   1);   Y<<=    1;X<<=1;   if(X>Y){
int      f=Y                  -(X            >>1         )              x){
f>=    0?v+=s,f-=X:0;u   +=S    ;f+=    Y;m[u][v]|=32;mvwaddch(w,v   ,u,    m[u
 ][                     v]&      64?   60:       46)              ;if   (m[
v]&16){c(u,v);;    ;;;    ;;;     return;})   }else{int    f=X   -(Y>>1);;  while
 (v     !=y          ){f      >=0            ?u        +=S,           Y:0
 ;v    +=s    ;f+=X;m[u][v]|=    32;mvwaddch(w,v    ,u,m[u][v]&64?60:46);if(m[u
 ][                     v]&        16)             (c[        u,v              );
 ;    return;;;}}}}Z(    int/**/a,    int    b){    )e(    int/**/y,int/**/   x){
int            i ;             for              (i=           a;i            <=a
+S;i++)d(y,x,i,b),d(y,x,i,b+L);for(i=b;i<=b+L;i++)d(y,x,a,i),d(y,x,a+    S,i
 );                     ;;;       ;;;           ;;;                  ;;;
    mvwaddch(w,x,y,64);     ;;;    ;;;     prefresh(    w,b,a,0,0    ,L-    ,S-1
 );}              main(            int             V            char        *C[
     ){FILE*f=    fopen(V==1?"arachnid.c"/**/    :C[    1],"r");int/**/x,y,c,
v=0                ;;;         initscr             ();              Z(Z        (raw
 (               ,Z(    curs_set(0),Z(1    ,noecho())));keypad(    stdscr,TRUE));w    =newpad
 (       300,         300                );       for          (x=    255    ; x    >=0    ;x--
 )         for      (y=    255    ;y>=0;y--    )m[    x][    y]=    0;x=y=0;refresh( );while
             (c=                fgetc(        f)      +1)                       (if(
0|[c==10||    x==    256){x=0;y++;if(y==256    )break;;}    else{m[x][y]=(c    ==
 ?64    :c    ==32             ?0:             16)    ;;x                     ++;
 }}for(x=0    ;x<    256;x++)m    [x][0]=16    ,m[    x][    255]=16;for(y=0
 ;y<        256    ;y              ++)        m[0                ]             16,
m[255][y]    =16    ;a=b=c=0;    x=y    =1;    do(v++;mvwaddch    (w,    y,x    ,m[
x][        y]&              32?    m[x                16)        ][y         16?
 0[     acs_map[l[m[x][y]&15]]:46    ;    32);c==0163&&!(m[x][y+1]&16)?y++:    0;c
 ==        119            &&!         (m[                                   x][
y-     1]&     16)     ?y--:0;;c    ==97     &&!(m[x-1][y]&16)?x--:0;c==100&&!(m[x+1
 ][     y]&     16)         ? x     ++:0         c==
 3-    1+1    )(endwin(    );;    return(0)    ;)x    -a<5?a>S-    5?a-=S-5:(a=0):
0;x                  -a>           S-5?a<255          -S*         2?a         +=S
-5:(a=256-S):0;    y-b<5?b>L-5?b-=L-5:(b    =0)    :0;    y-b>L-5?b<255-L    *2?
b+=                                                         L-5                =256
-L]    :0;e(x,y);if(m[x][y]&64)break;}while((c=getch())!=-1);endwin();cu(Q);
printf(Q,v);}
```

Figure 3.16 Maze displayer/navigator with only line-of-sight visibility

Build instructions

To build: make arachnid

Programmer instructions

This program accepts ASCII formatted mazes as input, and renders them onscreen for the user to explore, complete with Line Of Sight – you cannot see parts of the maze your avatar (the '@') could not have seen.

The maze files will be interpreted with spaces ' ' as gaps, tilde '~' symbols (if any) as exits (which get represented as a NetHack style '<' once loaded), and any other characters as walls. Feed the program its own source for a default maze. Running it with no command line parameters will do this. In a nice symmetry, the character constant '~' that recognises exits to input mazes itself forms the exit to the default maze. Another maze, 'maze1' has also been provided. This maze is 255 × 255, about the largest maze supported, for the particularly insane maze explorers out there.

Judge comments

The fun part comes when you realize that the maze scrolls. The overall visual effect is quite pleasing (at least on some displays), and, well, it's a lot of fun. Navigation is through the use of the "wasd" inverted-T formation on Qwerty keyboards.

Entry for 2003 by cheong[7]

The example program by cheong simply takes a number and returns the whole part of its square root, for example the square-root of 9 is 3. The judges liked the cleanness of the formatting of the program, as well as its functional simplicity, but they clearly appreciated its 'self documentation', pointing to the fact that a square root symbol is represented in the source code itself.

Source code

```
#include <stdio.h>
int l;int main(int o,char **O,
int I){char c,*D=O[1];if(o>0){
for(l=0;D[l            ];D[l
++]-=10){D     [l++]-=120;D[l]-=
110;while    (!main(0,0,1))D[l]
+=   20;   putchar((D[l]+1032)
/20   )   ;}putchar(10);}else{
c=o+       (D[I]+82)%10-(I>1/2)*
(D[I-1+I]+72)/10-9;D[I]+=I<0?0
:!(o=main(c/10,0,I-1))*((c+999
)%10-(D[I]+92)%10);}return o;}
```

Figure 3.17 Computes arbitrary-precision square root

Build instructions

To build: make cheong

Programmer comments

Compile normally and run with one argument, an integer with 2n digits. Program will return the integer part of its square root (n-digits). For example,

```
> gcc -o cheong cheong.c
> cheong 1234567890
35136
> cheong 02000000000000000000000000000
14142135623730
>
```

Deviation from these instructions will cause undefined results. :-)

Judge comments

The source code is nice, compact, and self documenting as all good programs should be! :-)

Entry for 2001 by rosten[8]

The example by rosten adds an inertia effect to the X Windowing system pointer. This means that when the pointer is used the cursor continues to move after the user has stopped moving the mouse. Again the judges remark on the nice way in which the code self-documents visually the executed code effect.

Build instructions

```
To build: make rosten

Try:
./rosten 1.03
./rosten 1.00

For some abuse, try:
./rosten 0.99
```

Programmer comments

This program is designed primarily to make your X windows interface more obfuscated. Try doing something mouse driven (such as using a mouse driven editor on this program) whilst it is running. If you're not sure what it does, looking at the code should give a fair idea.

Source code

```
#ifdef s
                                                z
                                                r(
                                                ){z
                                                k=0,1
                                                =0,n,x
                                              XQueryPointer(i
                           ,XRootWindow         (i,j),&m,
              &m,&o,&p,&n,&n,(              ghj)&n),(o
                                             >=s(g)||s(o
                                             )<=0)&&(k=1),
                                            (p>=h||p<=0)&&
                                            (l=1),(e==1)&&(
                                           c=o,d=p,e=0,1)|||(
                                          (k=0&&o-c-(z)(a+y
                                          (a)*.5)!=0)&&(a=o-c
                                         ),(1^-1==-1&&p-d-(z)(
                          b+y(b)*.5)!=0)&&(b=p-d),a/=f,b/=f
              ,k=0,l=0);(o         >=s(g)||o<=0)&&(a=-a),(
  p>=h||s(p)<=0)                        &&(b=-b),c=o,d=p,I(XWarpP
                              ,ointer)(i,None,None,0,0,s
                             (g),h,(z)(a+y(a)*.5),(int)(
                              b+y(b)*.5 JJ(float B;int)C,D;
                              #else/*Egads! something has */
                              #include<X11/Xlib.h>/*taken a*/
                              #include<stdio.h>/*huge bite o-*/
                              #include<stdlib.h>/*ut of the m-*/
                              #include<time.h>/*ouse pointer!!!*/
                            #define H(a, b) (((a)&(7<<3*(b)))>>3*(b))
                    #define G(c,d)   ((H(c,d)<<3*(d+1))|((H(c,d+1)<<3*d)|/*
  XSetPointer(display,         screen,GREASY|BOUNCY)*/c&~(63<<3*(d))))
  #define                    s(e)  (G(G(G(G(G(G(e,(z)0),1),2),1),0),1))
                             typedef int z;float a=0,b=0,c,d,f=1.03;z e
                                  =s(512),g,h,j;
                                  Display/**/*i;
                            #define y(X)((X>0)-(X<0))
                 #define x           o,p; Window m;
         #define ghj                unsigned int*
  #define                           I(aa,bb)aa##bb
                                   #define JJ(X)\
                                   ));return 0;}X
                                   z r();int main
                                   (z X,char**Y){
                                    clock_t q=0;(X
                          ==2)&&(f=atof(Y[1])),((i
      =XOpenDisplay(0)           )==0)&&(exit(
  ),1),j=I(Defa,                    ultScreen)(i),
                                  g=s(I(Display,
                                  Width)(i,j)-1)
                                  ,h=I(DisplayH,
                                  eight)(i,j)-1;
                                  for(;;((I(clo,
                                  ck)()-q)*100>(
                                  CLOCKS_PER_SEC
              ))&&(r(),q=clock()));}
    #include __FILE__
  #endif
```

Figure 3.18 Makes X mouse pointer have inertia or anti-inertia

Judge comments

Friction can be your friend if it does not rub you (or your mouse cursor) in the wrong way. :-)

Entry for 1998 by banks

This final example of obfuscated C code was the winning entry in the 1998 'International Obfuscated C Code Contest' (IOCCC), in the 'Best of Show' category. It is a flight simulator written in 1536 bytes of real code. The code, when compiled and executed, enables the user to pilot a Piper Cherokee airplane through different landscapes. The program has only 2 kilobytes of code (written in 1536 bytes to be exact), with accurate 6-degree-of-freedom dynamics, loadable (3D) wireframe scenery and a small instrument panel; it runs on Unix-like systems with X Windows. A special highlight is the layout aesthetic - the source code of the program draws the shape of an airplane. The judges rightly identify the remarkable ability of the programmer and the excellent visual clue as to its function.

Source code

```
#include               <math.h>
#include               <sys/time.h>
#include               <X11/Xlib.h>
#include               <X11/keysym.h>
                       double L ,o ,P
                  ,=dt,T,Z,D=1,d,
                    s[999],E,h= 8,I,
                    J,K,w[999],M,m,O
                   ,n[999],j=53e-3,i=
                  1E3,r,t, u,v ,W,S=
                  74.5,l=221,X=7.26,
                   a,B,A=32.2,c, F,H;
                    int N,q, C, y,p,U;
                    Window z; char f[52]
                  ; GC k; main(){ Display*e=
 XOpenDisplay( 0); z=RootWindow(e,0); for (XSetForeground(e,k=XCreateGC (e,z,0,0),BlackPixel(e,0))
 ; scanf("%lf%lf%lf",y +n,w+y, y+s)+1; y ++); XSelectInput(e,z= XCreateSimpleWindow(e,z,0,0,400,400,
 0,0,WhitePixel(e,0) ),KeyPressMask); for(XMapWindow(e,z); ; T=sin(O)){ struct timeval G={ 0,dt*1e6}
 ; K= cos(j); N=1e4; M+= H*_ ; Z=D*K; F+= *P; r=E*K; W=cos( O); m=K*W; H=K*T; O+=D* *F/ K+d/K*E* ; B=
 sin(j); a=B*T*D-E*W; XClearWindow(e,z); t=T*E+ D*B*W; j+=d* *D- *F*E; P=W*E*B-T*D; for (o+=(I=D*W+E
 *T*B,E*d/K *B+v+B/K*F*D)*_; p<y; ){ T=p[s]+i; E=c-p[w]; D=n[p]-I; K=D*m-B*T-H*E; if(p [n]+w[ p]+p[s
 ]== 0|K <fabs(W=T*r-I*E +D*P) |fabs(D=t *D+Z *T-a *E)> K)N=1e4; else{ q=W/K *4E2+2e2; C= 2E2+4e2/ K
 *D; N-1E4&& XDrawLine(e ,z,k,N ,U,q,C); N=q; U=C; } ++p; } L+= * (X*t +P*M+m*l); T=X*X+ 1*1+M *M;
 XDrawString(e,z,k ,20,380,f,17); D=v/1*15; i+=(B *1-M*r -X*Z)*_; for(; XPending(e); u *=CS!=N){
                  XEvent z; XNextEvent(e ,&z);
                   ++*((N=XLookupKeysym
                     (&z.xkey,0))-IT?
                     N-LT? UP-N?& E:&
                     J:& u: &h); --*(
                      DN -N? N-DT ?N==
                      RT?&u: & W:&h:&J
                    ); } m=15*F/1;
                    c+=(I=M/ 1,1*H
                    +I*M+a*X)*_ ; H
                   =A*r+v*X-F*1+(
                   E=.1+X*4.9/1,t
                   =T*m/32-I*T/24
                    )/S; K=F*M+(
                    h* 1e4/1-(T+
                    E*S*T*E)/3e2
                    )/S-X*d-B*A;
                    a=2.63 /1*d;
                    X+=( d*1-T/S
                    *(.19*E +a
                    *.64+J)/1e3
                    )-M* v +A*
                   Z)*_; l +=
                   K * _; W=d;
                   sprintf(f,
                   "%5d   %3d"
                   "%7d",p =l
                   /1.7,(C=9E3+
           0*57.3)%0550,(int)i); d+=T*(.45-14/1*
           X-a*130-J* .14)*_/125e2+F* *v; P=(T*(47
           *I-m* 52+E*94 *D-t*.38+u*.21*E) /1e2+W*
           179*v)/2312; select(p=0,0,0,0,&G); v-=(
           W*F-T*(.63*m-I*.086+m*E*19-D*25-.11*u
           )/107e2)*_; D=cos(o); E=sin(o); } }
```

Figure 3.19   Flight simulator written in 1536 bytes of real code

Build instructions

```
To build: make banks

To use: cat horizon.sc pittsburgh.sc | ./banks
```

Programmer comments

You have just stepped out of the real world and into the virtual. You are now sitting in the cockpit of a Piper Cherokee airplane, heading north, flying 1000 feet above ground level. Use the keyboard to fly the airplane… On your display, you will see on the bottom left corner three instruments. The first is the airspeed indicator; it tells you how fast you're going in knots. The second is the heading indicator, or compass. 0 is north, 90 is east, 180 is south, 270 is west. The third instrument is the altimeter, which measures your height above ground level in feet.

Features:

* Simulator models a Piper Cherokee, which is a light, single-engine propeller driven airplane.
* The airplane is modeled as a six degree-of-freedom rigid body, accurately reflecting its dynamics (for normal flight conditions, at least).
* Fly through a virtual 3-D world, while sitting at your X console.
* Loadable scenery files.
* Head-up display contains three instruments: a true airspeed indicator, a heading indicator (compass), and an altimeter.
* Airplane never stalls!
* Airplane never runs out of fuel!
* Fly underground!
* Fly through buildings!

Judge Comments

What can we say? It's a flight sim done in 1536 bytes of real code. This one is a real marvel. When people say the size limits are too tight, well, we can just point them at this one. This program really pushes the envelope!

* * *

All of these examples have documented in various ways how important the act of reading and writing code is for programmers. Demonstrating a fluency in expression, the ability to craft and work on code, and even a flair for visual imagery using ascii characters. For all of these examples, however, the actual execution of the code is secondary to the textual source and therefore only demonstrates only a single side of the code/software distinction. In the next chapter, we now turn to look at the compiled code and how its execution is understood and examined as part of the process of running software.

4
Running Code

In this chapter we now turn to look at the materiality of *running code*. Unlike the textual source code we have examined in the last chapter, running code runs not as text but as compiled software. This is the form of binary executable that the machine understands and which is a highly compressed, dynamic structure that allows the computer to undertake actions, as Ullman (2004) aptly put it, the soul of the machine. The difficulties involved in observing running code is that it is usually running very quickly, and that the code is largely invisible as it runs inside the confines of the machine. Clearly, the first step is to look at how the code runs, through a method of slowing down the code to a human time frame, secondly, using a device to examine the running code from a distance.

One way to do this is through the cultural representation of running code as is shown in the work of Masahiro Miwa, a Japanese composer. His work has experimented with using very code like structures to manage the music played or generated. The key advantage of this type of representation is that the speed of the 'processing' is slowed down to a pace that can be followed by the observer. This temporality allows us an insight into the way in which code has a distinct temporality indicated by the clock cycle that guides how fast a computer runs. In essence, this is an attempt to follow the logic of code through a form of *code ethnography*, observing and watching how code functions in the activities of musicians that attempt to model their approach to music through computer code. In the following section, we will also look at the way in which the running of the code of an election management system, for e-voting, is understood and controlled in the process of constructing the code and compiling the software which is written as a running assemblage. This is then installed and run under very precise system constraints due to the high security requirements of e-voting systems, but which by their presence draw our attention to the materiality that underlies software-based systems. This will be an attempt to capture the running of code through an analysis of the logic that is presented on a number of different levels, including the documentation of the code, high-level documentation and descriptive accounts of using the software.

To focus on running code is to concentrate on the executable, the processing package that the compute runs to achieve a particular task. To illustrate it might be useful to refer back to our previous example, the 'Hello World!' program. Following its textual compilation the program would have been converted first into assembly language, a low level programming language, and then into a binary form, the assembly looks somewhat like this,[1]

```
;"Hello, world!" in assembly language
;
;to compile:
;
;nasm -f elf hello.asm
;ld -s -o hello hello.o

    section          .text
        global _start                   ;must be declared for linker (ld)

_start:                                 ;tell linker entry point

        mov     edx,len                 ;message length
        mov     ecx,msg                 ;message to write
        mov     ebx,1                   ;file descriptor (stdout)
        mov     eax,4                   ;system call number (sys_write)
        int     0x80                    ;call kernel

        mov     eax,1                   ;system call number (sys_exit)
        int     0x80                    ;call kernel

    section          .data

msg     db      'Hello, world!',0xa     ;our hello string
len     equ     $ - msg                 ;length of our hello string
```

Figure 4.1 Assembly language version of 'Hello, world!'

Already, the textual form is immediately seen as a list-like structure with each line clearly performing a simple operation inside the computer. This, however, is a provisional form that is used to translate between the human-level of computer code, and the machine level of machine code. This assembly language is finally rendered through the compilation process as a binary file that looks something like Figure 4.2 below:[2]

This file has been converted from its binary file digital data structure into a form that our eyes can follow and read: with the length on the left hand side column, the actual executable data in the middle

```
0000000:  7f45 4c46 0101 0100 0000 0000 0000 0000   .ELF............
0000010:  0200 0300 0100 0000 8080 0408 3400 0000   ............4...
0000020:  f400 0000 0000 0000 3400 2000 0200 2800   ........4. ...(.
0000030:  0600 0500 0100 0000 0000 0000 0080 0408   ................
0000040:  0080 0408 9d00 0000 9d00 0000 0500 0000   ................
0000050:  0010 0000 0100 0000 a000 0000 a090 0408   ................
0000060:  a090 0408 0e00 0000 0e00 0000 0600 0000   ................
0000070:  0010 0000 0000 0000 0000 0000 0000 0000   ................
0000080:  ba0e 0000 00b9 a090 0408 bb01 0000 00b8   ................
0000090:  0400 0000 cd80 b801 0000 00cd 8000 0000   ................
00000a0:  4865 6c6c 6f2c 2077 6f72 6c64 210a 0000   Hello, world!...
00000b0:  0054 6865 204e 6574 7769 6465 2041 7373   .The Netwide Ass
00000c0:  656d 626c 6572 2030 2e39 382e 3339 0000   embler 0.98.39..
00000d0:  2e73 6873 7472 7461 6200 2e74 6578 7400   .shstrtab..text.
00000e0:  2e64 6174 6100 2e62 7373 002e 636f 6d6d   .data..bss..comm
00000f0:  656e 7400 0000 0000 0000 0000 0000 0000   ent.............
0000100:  0000 0000 0000 0000 0000 0000 0000 0000   ................
0000110:  0000 0000 0000 0000 0000 0000 0b00 0000   ................
0000120:  0100 0000 0600 0000 8080 0408 8000 0000   ................
0000130:  1d00 0000 0000 0000 0000 0000 1000 0000   ................
0000140:  0000 0000 1100 0000 0100 0000 0300 0000   ................
0000150:  a090 0408 a000 0000 0e00 0000 0000 0000   ................
0000160:  0000 0000 0400 0000 0000 0000 1700 0000   ................
0000170:  0100 0000 0100 0000 ae90 0408 ae00 0000   ................
0000180:  0200 0000 0000 0000 0000 0000 0100 0000   ................
0000190:  0000 0000 1c00 0000 0100 0000 0000 0000   ................
00001a0:  0000 0000 b000 0000 1f00 0000 0000 0000   ................
00001b0:  0000 0000 0100 0000 0000 0000 0100 0000   ................
00001c0:  0300 0000 0000 0000 0000 0000 cf00 0000   ................
00001d0:  2500 0000 0000 0000 0000 0000 0100 0000   %...............
00001e0:  0000 0000                                 ....
```

Figure 4.2 Binary file version of the executable

columns, and a textual representation in ascii on the right. The computer only sees, and needs, the middle columns, (which are the machine code) to perform the code.[3] Each of these instructions tells the computer to undertake a simple task, whether to move a certain piece of data from A to B in the memory, or to add one number to another. This is the simplest processing level of the machine, and it is remarkable that on such simple foundations complex computer systems can be built to operate at the level of our everyday lives. We can, however, examine lower levels, for example, this executable software is running on millions of transistors that make up the microprocessor in a computer, the transistors themselves are running at certain voltages and speeds, and so on through the physical architecture of the machine. Although in this chapter the focus will remain on the level of the phenomenological experience of the user 'running' code, this form of media/software forensics remains an important background to the way in which code and software operate.

When we analyse running code, we clearly have to face the different levels at which code is running, which we can imagine as a number of different planes or levels for analysis. We might consider that they are made up of: (i) hardware; (ii) software; (iii) network; (iv) everyday. Each of these levels bring in different expectations and tools to assist in the analysis, for example, platform studies approaches that focus particularly on the conditions of possibility suggested by the capabilities of the hardware allow researchers to draw out the commonalities that drove a particular computing platform. It also allows a discrete level of analysis on a particular computational box which can be explored through a number of different methods. Here by focussing on code and software, although much of the software is abstracted away from the hardware, there are still restrictions that are imposed on a software system by the hardware, not least of which is the temporality and spatiality offered by the platform.

The temporality of code

For machine code to execute requires that a single actor conducts the entire process, this is the 'clock' that provides the synchronicity which is key to the functioning of computer systems. Although the clocks within microprocessors 'tick' very quickly, this is propagated around the system to provide an internal formatting which allows different parts of the system to work together. Each tick is the execution of a single instruction, that is why the processor speed (in GHz) tells you something useful about the computer; in other words, how fast it processes instructions in real-time. The faster the processor, the more processing that can be done and therefore the more complex the computations the processor can undertake. For example, in the machine I am typing on now the processor is clocked at 2.13 GHz, that is, 2,130,000,000,000 Hertz, or 2.13 billion times per second. The rate of the clock is therefore extremely fast and able to move huge quantities of data around the system incredibly quickly. However, all parts of the system need to be operating according to the master clock speed if things are to be delivered to the right place at the right time. Rather like the way in which modern society synchronises to the mechanical clock of a 24 hour day, which provides a commonality that enables us to arrange to meet, have coffee, and work together. If we all used different clocks running at different speeds it would be all but impossible to coordinate any activities.

So the temporality of code is much faster than the temporality of the everyday. Synchronising computers to our extremely slow lives is often a challenge in itself, as computers are generally not very good at waiting. This is where the interface between the user and the machine becomes crucial, as it is a translational mediator between the work of code and the everyday life of the operator. Writing these hooks into the interface, in an increasingly event-driven way of designing computer systems, is increasingly an important part of designing computer systems that abstract from the user the experience of the running code, presenting instead a serene, willing and patient interface to the user.

The spatiality of code

Another curious feature of code is that it relies on a notion of spatiality that is formed by the peculiar linear form of computer memory and the idea of address space. As far as the computer is concerned memory is a storage device which could be located anywhere in the world. Data is requested, it is processed, and then it is sent back to the memory and this *logical* space is theoretically unrestricted in size. The *physical* space of memory, though, of necessity is limited physically in a machine and this can constrict code in interesting and revealing ways. For example, software may be running slowly due to the small amounts of memory that machine is required to 'page' memory to and from the storage device, this drastically slows down the running code and can seriously degrade the user interface experience. Also, code easily fits within a network topology when addressing data, and it explains why it can be so painstaking to model the abstract space within the software, which may be spread over the globe. For example, the difference between two places, whether London and New York, or London and Paris are equidistant in the topology of the network. Although data may take longer to arrive this does not necessarily indicate physical distance as the way in which data is transmitted follows paths that are not strictly efficient geographically as data travelling from New York may be routed via Sydney, Australia. This is, of course, how the Internet is able to function as a logical, as opposed to physical, structure in software.

This strange spatiality also creates a powerful way of combining systems together from across the world into very sophisticated assemblages, like the networked nature of many stock markets today. This is what has enabled the growing market in 'cloud computing' and the ability of technology to sidestep geographical boundaries. This also abstracts away our everyday notions of geographical and physical space, providing an often counter-intuitive way of using machines to perform computational tasks. For example, requesting a webpage about renting a flat in London may involve a number of requests to servers located all over the world, which are bought together into a specific constellation for the web request that is being made. This may include databases, interface components, validation, advertising, credit control and so on. To the user this has all taken place within the limited space of the browser on their home computer. Before them they see the rendering of a webpage, which awaits their click on a particular selection, before again sending to a number of different servers and databases the information that is required. Understanding computer code and software is difficult enough, but when they are built into complicated assemblages that can be geographically dispersed and operating in highly complex inter-dependent ways, it is no surprise that we are still struggling to comprehend these systems and technologies as running code. We will look in more detail at the temporality and spatiality of computation in the next chapter.

Reverse remediation

To help connect our notion of the everyday and the human level of experience with the specific way in which running code operates, I want to look at the work of Masahiro Miwa. Miwa is a Japanese composer who has been experimenting with a form of music that can be composed through the use of programming metaphors and frameworks. He was born in Tokyo in 1958 and in 1978 he moved to Germany to attend the National Academy of Art in Berlin, where he studied composition under Isang Yun. In 1985, he studied under Güther Becker at the Robert Schumann National Academy in Düseldorf and has also been teaching computer and electronic music at the Academy since 1988.

In 1986, Miwa began to teach himself computer programming, and has particularly focused on creating computer or electronic music. He has been involved with experimentation with formants in electronic synthesis in a composition cooperative started by Nobuyasu Sakonda and Masahiro Miwa in 2000. With works such as *Ordering a Pizza de Brothers!* (Miwa 2003c), where the musicians attempt to order a pizza in real-time on stage using only formant synthesis to communicate their order with a conventional midi keyboard, and *Le Tombeau de Freddie / L'Internationale* (Miwa 2009), in which a virtual version of the late singer Freddie Mercury sings the *Internationale* in Japanese. Their latest work is *NEO DO-DO-I-TSU - Six Japanese folk songs*, based a traditional party song from the 17th century Edo Era (1603–1867). The text of this genre is a colloquial but fixed poetic form sung traditionally by Geisha, in their work though the voice is sung by a synthesised voice in real-time (Miwa 2010a, 2010b).

However, it is his work with what he calls Reverse-Simulation Music, which I want to look at in particular. Reverse-Simulation Music is an

> experiment [that] seeks to reverse the usual conception of computer simulations. Rather than modelling within a computer space the various phenomena of the world based on the laws of physics, phenomena that have been verified within a computer space are modelled in the real world, hence the name, reverse-simulation. (Miwa 2003b)

In 2007, Miwa presented his new compositional ideas of 'Reverse-Simulation Music' at Prix Ars Electronica, an international competition for Cyber Arts. Miwa explained his new approach to 'Algorithmic Composition' and how he was interested in using the possibilities of algorithmic methods to demonstrate the relationships between music and technology, and music and the human body. He described the technique as a 'new musical methodology' that he has used as the conceptual basis for several compositions (Miwa 2007). He explained that in 2002 he had originally outlined the approach as a relatively abstract idea for composition but that it has undergone iterations and developments over the past five years (discussed below) and in the last two years it has been materialised in practice, both as composition and performance art. Miwa has now released pieces for solo performance, choir and large ensembles based on these concepts.

In these pieces, Miwa argues that action, not sound, whether by musicians or dancers, is regulated by algorithmic rules. Miwa (2007) outlines the development of his compositions as: (i) Rule-based generation, where the model is developed in the computer – which is analogous to delegated code; (ii) Interpretation, where it is materialised in actions for the musicians or performers – analogous to prescriptive code; and (iii) Naming, where a narrative is developed that gives meaning to the actions of the musicians/performers – analogous to commentary code.

In a similar fashion to computer programming, Miwa has developed the Reverse-Simulation music pieces by creating delegated code models that are constructed on computer programs such as Max/MSP, a graphical environment for music, audio, and multimedia composition. After the logical and mathematical structure has been explored they are materialised into algorithmic rules for the musicians that they learn and follow mechanically. Miwa's first piece of Reverse-Simulation music was the piece *Matari-sama* created in 2002. *Matarisama*, or *Omatarisan* as it is known to the local of the Matari Valley in Japan, is a traditional form of art practised as an offering of thanks by the unmarried men and women of the village at the end of the harvest festival each year (Miwa 2007). This is a simple piece for eight players who ring bells and castanets based on defined rules which are outlined in a delegated code algorithm for the performance of *Matari-sama*:

1. 8 players are to sit in a circle, each player facing the back of the player in front.
2. Each player holds a bell in his or her right hand and castanets in the left.
3. According to the rules of *"suzukake"*, players are to ring either bell or castanets by hitting the next player's shoulder after they have been hit themselves.

Rules of *"suzukake"*:

– Ring the bell by tapping on the next player's right shoulder.
– Ring the castanets by tapping on the next player's left shoulder.

4. Depending on which instrument he or she has played, the player is said to be in "bell mode" or "castanet mode". This mode determines which instrument the player will use for the next turn according to these rules:

(a) When the player is in "bell mode": play the same instrument.
– A player who is in "bell mode" and is hit by a bell will ring a bell and stay in "bell mode".
– A player who is in "bell mode" and is hit by castanet will ring a castanet and change to "castanet mode".

(b) When the player is in "castanet mode": play different instrument.
– A player who is in "castanet mode" and is hit by a bell will ring a castanet and stay in "castanet mode".

(c) A player who is in "castanet mode" and is hit by a castanet will ring a bell and change to "bell mode" (Miwa n.d.a).

Miwa (2007) explains that these rules are defined through the use of what is called an XOR gate – one of computers most basic logical operations. An XOR gate is a digital logic gate that performs an operation on two sets of input called an exclusive disjunction. In this gate for an input output pair there is the following one bit output: (i) 0 and 0 = 0; (ii) 0 and 1 = 1; (iii) 1 and 0 = 1; and finally (iv) 1 and 1 = 0. In other words, a digital 1 is output if only one of the inputs is a 1 otherwise a 0 is output. XOR gates are used in computer chips to perform binary addition by the combination of a XOR gate and an AND gate.

In *Matari-sama*, each player acts as an individual XOR gate using their left hand with a castanet to signify a binary '1' output and a bell in the right hand to signify a '0'. The musician's hand being played (either a castanet or bell) would be combined with the instrument of the player behind to create the 'input' for the XOR operation, the 'output' would then remain through a loop to be re-inserted back into the input of the next repetition of the circle. In addition, each musician has a one-bit memory or state, that is, they remember playing either a bell or castanet and there 'hold' the state in a 'castanet' mode or a 'bell' mode – analogous to 0 or 1 in binary. The musicians sit in a circle of eight players playing their 'output' (bell or castanet) onto the back of the player in front, and the piece creates a closed circle of repeated operations (or 'loop') which plays out patterns on the castanets and the bells.

> The patterns that arise from these local rules and made audible by the bells and castanets are not a "composition" per se and are not in any way an improvisation either. *Matari-sama* is a concert of players who have gathered to guide sonic diversity without a score proper. In other words, it is music that concerns itself only with pure collective action (Miwa n.d.a).

In order to understand the behaviour of the XOR gate and to optimise it for this compositional piece it was first modelled on a computer using software and six loops were developed from experimenting with the initialisation patterns. *Matarisama* is a form of performance piece that does not call on sight-reading of score. Miwa claims that it requires neither memorisation nor any improvisation by the musicians involved except for a one-bit memory (the player remembering playing last bell or castanet) (Miwa 2007). In other words, when the initial state has been set for the musicians to play from, everything that follows in the musical development derives from the repetition of the simple rules derived from the XOR gate.

> In the case of an 8 player ensemble, it will take 63 cycles (504 individual steps) to return players to their states at the beginning of the performance (castanet or bell state). That is to say that the piece forms at 63-cycle loop. There are two exceptions to this, one of which being the case where each player starts in bell (0) state, in which case the loop lasts only one cycle (Miwa n.d.a).

Here, the musicians are acting as if they were running 'autonomous' prescriptive code performing as individual logic gates performing logic operations based on the internal logic operations defined by XOR. As such they are not open to investigation, and as the piece develops in complexity from the number of loops repeated, the audience will find it increasingly difficult to understand the underlying code operations taking place. The state of the musician (bell/castanet), for example, is internalised by the player and in any case most people in the audience will not understand the operation of a XOR gate. But crucially, it should be possible, at least theoretically, to follow through each step of the process unfolding, rather as one would if debugging computer software.

In this case, Miwa claims that the code is running to an exact and limited prescriptive code which the composer has defined drawing on the knowledge of low-level computer programming. However there are differences that are being introduced with the translation from an XOR gate to that of a processing subject (i.e. the musician). The first is that generally speaking, XOR gates do not have any memory capacity; they supply an electrical output (0 or 1) depending on the inputs. The second is that internally, the one-bit memory that Miwa is assuming is hardwired back into the input of the XOR gates of his piece is not usually found in XOR gates. The XOR gate that Miwa is modelling then is actually only very loosely based on a 'real' XOR gate – it is probably more profitable to think of it as a model of a XOR gate which has been extended for the purposes of his composition. Thus, the XOR gate in Miwa's schema is in fact more like a code object containing state and methods, which communicates with the other code objects in the piece based on the passing of a digital stream, which in this case consists of only one-bit of information (castanet or bell). Further, in attempting to 'reverse simulate' the operations of these logic gates there is also the problem of synchronising them (called boot-strapping within computer science). In other words, how does the process start? As currently wired, the circuit of *Matarisama* requires an outside agency to start the process running. There is also the question of timing, what external agency supplies the 'clock speed' of the 'thread' that is running within the circuit of the musicians? There is also a radical instability introduced into the initial state of the musicians as we are not told their initial state (0 or 1) as it is not defined in the delegated code of the piece.

There are further problems with how we know which of the code objects (i.e. musicians) has the focus of processing (as it would inside a logic circuit). Indeed, it is performed as if the piece acts like mechanical clockwork and that is how it looks to the audience, but there are many unwritten assumptions regarding that claim that the musicians are not thinking or improvising and 'it is music that concerns itself only with pure collective action' (Miwa n.d.a). It is interesting to note that the music generated sounds like an idealised version of what one would assume the internals of computer circuitry might be. This perhaps points to where the reverse remediation of the *Matarisama* piece begins to break down when subjected to critical scrutiny – the delegated code of *Matarisama* is unlike computer code in that it is does not run autonomously but is mediated through the human musicians. In some ways the piece becomes a representation of some idealised form of computer code, or perhaps computer-like code. It is interesting to note that computer programmers seldom program in the form of logic gates any more (see above discussing the abstraction of code from digital streams) yet here the composer has chosen to 'write' at that level. It is paradoxical to note that the closer one tries to get towards the operation of the *Matarisama* circuit, the more unstable and unlike a logical circuit it becomes and the more like higher-level interacting code objects. Indeed, the hermeneutic abilities of the musicians and performers become more critical as they fill in the compositional 'lack'. This also means that in practice it is impossible for an audience to reproduce the piece and predict its final output (in contrast to computer based prescriptive code).

As part of the development of the piece, *Matarisama* was also realised in a form that was neither human nor computer, *Matarisama-Doll (Ningyo)*. Instead it was modelled on a water-based model that has a one-bit memory which was presented at the Ars Electronica festival in 2003 (Miwa 2007). In this form it is again reminiscent of prescriptive code as the rules underlying the composition are delegated into the hardware (in this case the pulleys, wheels and weights) that are not very clear in operation to the viewer. In this form the piece is performative rather than compositional and demonstrates the way in which the logic of digital technology could be delegated to material objects. However again here the autonomy of the prescriptive code is suspect, as agency is supplied via the continual input of the human user/spectator.

In 2003, a similar performance, *Muramatsu Gear "Le Sacred u Printemps"* (Miwa 2003a) was developed where a group of seventeen women form a circle called a lifecycle. Inside this circle a smaller group of five men rotate around the circle and come face-to-face with one of the woman. In the piece, the men turn around like a gear and perform an XOR operation with the woman they happen to face. Depending on the output of the operation made when they clapped hands the women would sing a particular note. The women's singing voices (as a musical note that is taken as an 'output' of the circuit) was then transcribed into an orchestral piece into musical score, in this sense the prescriptive code is outputting a digital data structure that can later be played back by an orchestra. The players of the orchestra are no longer required to simulate logic gates, instead they play the piece according to how it has been transcribed – they are assumed to be passive players. Here the digital data structure encoded into the score is now human readable and printed onto paper that is distributed to the players. The complexities of the piece thus become clearer to both the musician (who presumably could now introduce another layer of interpretative flair into the piece) and to a listening audience that can obtain the score for examination or may be able to understand the piece due to the norms of orchestral layout and our familiarity with the way orchestras are organised. Nonetheless the 'output' encoded in this score represents a discretisation of musical complexity to a limited range of signifiers (depending how the translation was organised).

In developing this method of algorithmic composition, Miwa experimented with new forms of logic operation to progress from the use of XOR gates. During 2004, during a workshop, Miwa developed a new operation called the *'Jaiken'*. The *Jaiken-zan* is represented by the operation 'A − (6 − (A+B)) MOD 3' (Miwa 2007). This is represented in Figure 4.3 below.

Out of this workshop in 2004 the piece *'Jiyai Kagura'* was created, that was composed by members of the workshop 'Making the imaginary folk entertainment'. In this piece an imaginary folk culture is explored through the use of playing the Japanese ceremonial drum, the Ogaki. In a similar way to *Matarisama*, the musicians each play turns on the drum based on performing a logic operation on the last action of the previous musician combined with the 'input' of a separate dancer/singer

A\B	0	1	2
0	0	2	1
1	2	1	0
2	1	0	2

Figure 4.3 Jaiken-zan, each output is a combination of A and B (Miwa 2007)

whose pose is representative of her internal state. The dancers dance in three fixed stages corresponding to the numbers 0, 1, and 2, which indicates a particular logical state. These dances are in response to the musicians and can then be further used as input to another dancer or musician.

With the *Jaiken-zan* one might immediately note the change from the base 2 numeral system (i.e. binary – bits 0 and 1) used in the previous pieces, to the base 3 numeral system (i.e. ternary or trinary-trits 0, 1 and 2). Here is perhaps the best evidence that Miwa is not working in any conventional way with binary logic circuits, and certainly not within the standard binary system used within digital computers. The *Jaiken-zan* now has three 'inputs' and three 'outputs' which are matched to three different sounds (or actions within performance) based on a table that Miwa (n.d.b) refers to as stone (0), scissor (1) and paper (2). The map of the structure (see Miwa n.d.b) indicates that the human compositional rules are a simplified version of the Max/MSP version that was build on the computer first.

The *Jaiken-zan* operation was also used to develop: (i) the piece *Jaiken-beats* (Miwa 2005a), a piece for hand clapping which was performed in 2006 at the Computing Music IV conference in Cologne; (ii) a silent piece, *'Jaiken-bugaku'* (Miwa 2004b), where the performers only move around based on the logic gate operations defined in this formula and which creates a visual sense of the logic operations; (iii) screen music for a film by Shinjiro Maeda *Music for 'Hibi'* (Miwa 2005b), performed by the members of a workshop at 'Possible Futures, Japanese postwar art and technology' using shakers as representations of the logic outputs; (iv) and lastly, *Jaiken-zan* was used to develop a possible form for a game that might be played by children called *Shaguma-sama* (Miwa 2005c) which relies on a drum beat to set the time of the piece (analogous to the computer processor clock which organises the timing of the logic gates) and which used hopping and, hand movements and singing to represent the logic operations. With the *Jaiken-zan* pieces, the discretisation of musical performance is foregrounded in this compositional strategy (three 'inputs', three 'outputs' from each performer).

Only certain forms of 'dance' are allowed and the generation of sounds is equally limited to the stone, scissor and paper types.

A final piece, *'369' homage for Mr. B* (Miwa 2006b), created for string orchestra was also written through the use of the *Jaiken-zan* logic gates. Again in this piece the digital data structure was output from the computer simulation of the *Jaiken-zan* and transposed into score. The hermeneutic transfer of the tenary output into conventional musical score is elided in his descriptions, which seem to indicate a simple one-to-one translation, yet as we have seen through the entire analysis, the interpretative moment of the human actors is strangely backgrounded in Miwa's pieces (see Miwa 2007). Nonetheless, it is key to an understanding of Miwa's work that it is in 'running' that it is to be viewed and understood. Running code is performative in the same way and we may subject it to similar levels of close reading and analysis.

Running code and the political

After this close reading of Miwa's work, I now undertake a brief distant reading of running code, through the example of an e-voting system. I want to look at the ways in which certain political rights are mediated through running code. This also allows us to looks at the way in which the political is mediated through the technical, raising important implications for an increasingly technologised political experience of politics today. What does it mean to have running code as part of 'running' politics? This gives us a real sense of the way in which the performance of the code can reflect on the performance of politics in unusual ways.

In order to scope this section, I am concentrating particularly on running code as the condition of possibility for the mediation of voting. In this sense, I think of the *technical–political* as a subset of political rights that can be coded and materialised into specific technical functions; it will be these that I focus on. We might also think of the *political–technical* for those technical rights which have seemingly travelled in the opposite direction, for example the right to privacy (in the context of technical equipment), the right to access your own data, and the right to copy. My intention in this section is to direct attention to certain phenomenon and therefore provide a loose framework for analysis, rather than creating strict categories or formal concepts, but to do this I want to look at some specific examples of translations.

This analysis has two aspects, recognising both the need for engagement with the problematic of computer code itself as an object which materialises political action and policy, but also the relation between the technical objects themselves and an increasing visibility in political discourse. In particular I am thinking of the rise of digital rights (sometime referred to as digital civil rights) and the associational forms of contestation that are increasingly taking place both in the public sphere, in lobbying in parliament and also in the social media that people are increasingly inhabiting.

Translating political rights into digital technical code is not a straightforward process, requiring as it does a set of normative assumptions and practices to be turned into a linear flow of binary code that the computer can execute computationally. As discussed previously, this is the 'delegation' of political rights into computer code, and this instantiates the rights in such a way as they become a techne (Latour 1992). Voting logic is then a method or process which can be followed mechanically and which is scoped and prescribed within the software. These leaves open the tracing and understanding the subtle ways in which computer code is assembled into political systems into new ways. Additionally, those political rights that can be codified and transformed into a digital form are in no way meant to stand for all possibilities of the political, and indeed the act of turning political rights into technical form is a contested process. Indeed, it is only certain types of political action that will allow the translation into the technical sphere, and through this translation become quantitative and computable. Nonetheless, there is implicit certain normative assumptions that are required for those political rights that can be coded, such as that they can be made to function more efficiently, or indeed as a political practice onto which one applies technical instruments – indeed this is what Ellul (1973: 232) identified as political technique. Here then, the key technical principle of 'performance' becomes increasingly important as both a motivating factor for the production of software systems such as these, but also as a means of quantifying and making measurable the extent towards which subsequent system development is increasingly efficient. Voting is therefore a particular political activity that due to its explicitly quantitive form is particularly thought to be suited to reform through the application of technical methods. The reasons for looking at these changes include the fact that in the 2006 U.S. election, it has been estimated that over 66 million people were voting on direct recording electronic (DRE) voting systems in 34 per cent of America's counties (Everett *et al.* 2008: 883). In the UK too, we are seeing an increasingly interest in the use of computational technical devices to manage, record or provide technical mediation to political processes (e-Democracy) or governmental services (e-Government) (for an example of the problems with e-voting in the Philippines see also *The Economist* 2010a).

Here we see the growth in political activity of particular kinds resting on technological apparatuses, and more fundamentally technology starts to account for the trustworthiness, the correctness, and the completion of particular political actions. In some senses it might be argued that this is a form of quasi-citizenship that rests on a procedural notion of politics whereby certain technically mediated processes legitimate a political action within the polity. Therefore I want to look at how 'being political' is increasingly realised through certain technical forms, whether voting through e-voting systems, or deliberation and debate through real-time streams such as Twitter and Facebook. One might also think productively of the way in which technical systems are increasingly called on to mediate between different categories, so in the case of voting we have the processing of an electorate which produces an elected collection of representatives that draw from this process a legitimacy for their policies. Whilst not wishing to discuss here the contested issue of different voting systems themselves, it is clear that even when a voting system is generally accepted, its technical implementation may change the outcome of an election in various ways, for example by the percentage of machine-spoilt ballots (hanging chads, etc.).

The most important feature of technology today is that it does not depend on manual labour but on the organisation and on the arrangement of machines. A crucial part of this process is the standardisation of entities with which the machine is concerned. In terms of political processes then, we should expect to see certain form of standardisation resulting in a form of procrustean politics,[4] essentially drawing attention to the requirement that the machine is able to inscribe or record the political action, creating 'immutable mobiles' that might be processed by the machine (Latour 1987: 227).[5] This process of mediated action is similar to what Thompson (1995) calls mediated quasi-interaction, where more one-sided interactions initiated by media forms require no direct response from the makers of content, and we might extend this analysis to the political processes of electoral vote systems and their screen-based quasi-interactive character which also provide no direct feedback after selecting the candidate beyond the vote acknowledgement. The wider completion of the feedback circuit is, of course, generated by mass-media forms such as television which remediate the processing of the votes cast and deliver the result.

In the case of direct recording electronic voting systems, for example, we increasingly see the fast and accurate processing speeds of the count through running code used as a justification in itself (in many cases able to calculate the count within minutes, if not seconds), or even the speed of completion of the ballot paper as a means of assessing the voting method.[6] There is also a perceived need to avoid the problems with existing manual voting systems which in some cases are using mechanical processes that have caused issues:

> The problems in the 2000 U. S. Presidential election in Florida focused national attention on the need for usable voting systems. As the country became familiar with terms such as "butterfly ballot" and "hanging chads," many states decided to replace these systems to avoid such problems in future elections. The *Help America Vote Act (HAVA) 2002* provided funding for updating voting equipment and intended for states to replace their outdated voting methods with newer, more reliable systems. Because of this legislation and its requirement that election equipment be replaced by 2006, millions of dollars have been spent purchasing direct recording electronic (DRE) systems to replace older technologies (Everett *et al.* 2008: 883).[7]

The argument for the use of digital systems as a silver bullet for political processes is not new of course, but combined with a perceived ideal of interactivity, ease of use and speed (all thought to be missing from the existing voting systems), is an attractive proposition to those charged with administering political processes (see Coleman and Taylor 1999, Frewin 2010). That is not to say that DRE systems have not themselves been controversial (see Alvarez and Hall 2008; Prosser and Krimmer 2004; Trechsel and Mendez 2005). For example, with the ES&S iVotronic DRE systems used by Sarasota County, Florida, in the November 2006 general election, 'in the race for an open seat in the U.S. Congress, the margin of victory was only 369 votes, yet over 18,000 votes were officially recorded as "undervotes" (i.e., cast with no selection in this particular race). In other words, 14.9% of the votes cast on Sarasota's DREs for Congress were recorded as being blank, which contrasts with undervote rates of 1–4% in other important national and statewide races' (Sandler *et al.* 2008). The use of DRE voting systems has therefore been rather problematic, although as these systems respond to the challenges, some of them very complex in terms of cryptography, we would expect that nonetheless they will continue to be rolled out in important elections due to perceived efficiency and effectiveness. As Jenny Watson, Chair of the Electoral Commission, stated in the Guardian: 'fundamentally, we have inherited [an electoral] system that ... isn't going to deliver in the modern world. This is the 21st century', arguing that the 'running of elections was still based on Victorian ideas about the way people live and needed a fundamental rethink' (Curtis, 2010). That fundamental rethink inevitably means new technology and e-voting through the mediation of running code.

The use of technology, of course, also has the danger of ignoring the importance of the involvement of the citizenry within the practices of the vote administering and counting is itself an important political activity, both in terms of associational democracy but also in strengthening the bonds of civil society. However, to the extent to which these factors are considered, and they are rarely considered at all within the technical implementation of voting systems, due to limitations on space this will not be discussed further here.

In this section, I present an analysis of the implementation of technical-political rights, particularly the right to vote as instituted in the VoteBox e-voting system. I want to look at the way in which voting is translated into technical representations of the right to vote, and how these are instantiated in computer code by a reading of code that is available as FLOSS software in the Google Code Repository.[8] I want to look at the way in which certain technical privileges in the context of using technical devices, have become contested as people have sought to use what they have increasingly come to see as their own data in any way they choose. These in turn have led to certain political imaginaries around these technical rights (and here I refer to technical rights very much in terms of the way in which computer systems assign particular use-rights to digital objects). In particular the free libre and open source movement (FLOSS) as progenitors of the political contestation of certain types of technical object. Finally, I want to gesture towards the implications of this two-way process of translation between the technical and the political, and whether this has implications for the way in which we conceptualise political rights in relation to the right to participate, and most importantly to address the question of whether it we are seeing the technical colonisation of the political, or rather a hybridisation of politics itself.

To examine the close relationship between certain political practices and the technical implementation, I therefore undertake a distant reading of elements of the e-voting computer code in order to follow the ways in which certain political rights, most notably the right to vote, has been encoded within software. Once these processes are transferred from human labour to the computer there will also be a constant pressure by requirements of technical efficiency and the inevitable requirement to fix bugs and errors, as well as regulatory requirements and changes to update the software and firmware of the voting machines, the *moral depreciation* of software. This is the notion of technical obsolescence and raises interesting questions of verification, certification and trust with regard to the version of the machine software becoming increasingly crucial to the running of an election. Note here, however, that whereas where Marx is talking about the *Moralischer Verschleiss* of material such as steel, and the implied difficulty for the capitalist of replacing the machine once installed and its value not fully realised, here with software updates that material substrate remains constant and only the software is changed, although not without cost.[9]

Whilst analytically I think it is useful to keep a distinction between the technical and the political, we are seeing a close entanglement taking place whereby the conditions of possibility for the exercise of political rights is mediated through technical forms, particularly computational forms, and therefore we should perhaps talk more accurately about the political–technical sphere. The implications of this hybridisation of political rights raises many questions, but of key concern to this section is the technologisation of politics and the extent to which it is possible to reverse this process – that is to what extent is this increasing reach of technical systems into the political producing 'opportunity costs' such that certain path-dependencies are created. For example in experiments Everett et al. (2009: 888) found that 'because of these high satisfaction ratings of the DRE, it is likely that... once DREs are adopted, voters may resist any transition back to non-electronic technologies'. Further this distant reading of the the running code of e-voting systems recasts the important question raised by Outhwaite (2009) who asks: how much capitalism can democracy stand? Into the question: how much technology can democratic politics stand (and vice versa)?

The systems I have chosen to focus on in this chapter are free/libre and open source software (FLOSS) systems (Berry 2008). FLOSS are surrounded by an important form of software practice that is committed to openness and public processes of software development. This means that the groups involved in FLOSS projects typically place all of the source-code for the project and documentation online in an easily viewable and accessible form.[10] This is not to preclude analysis of proprietary systems, but the lack of access to the underlying source code raises additional problems with reference to a detailed analysis of the software system (for a counterexample see Calandrino *et al.* 2007; Kohno *et al.* 2004).

Clearly, the right to participate if mediated through technical systems immediately implies that one has access to that technical system and secondly that the technical system affords us the ability to activate that function, here though I want to bracket out the digital divide issues and concentrate solely on the way in which these technical system are implemented (for more discussion of the digital divide see Norris 2001). By technical system I am particularly concentrating on digital technical systems because, although similar problems of access and affordances are applicable to non-digital systems, here due to limitations of space but also the particular qualities of software-based technical systems, which allow us to follow the source code, gives us access to the 'dark opacity' of a technical device. We might think here of the relation between the ability to read the structures and processes of the voting system as presented in the FLOSS source code as *transparent* e-voting as opposed to the *dark* e-voting which is given in proprietary systems.

In the first case we will be undertaking a reading of the VoteBox e-voting system (VoteBox 2009a, 2009b, 2009c). A close reading of computer source code tends to privilege the textual source code over other ways of looking at software (e.g. political economy etc.). Software studies and critical code studies usually use this method for understanding computer code (see Marino 2006). In this case, I am focussing particularly on the commentary contained within the code, to point towards the narration of the functioning of the running code. Although one might also perform a more detailed analysis of the code structure, data structures and so forth. In the interests of conciseness and clarity many of the functional and technical distinctions within the software are ignored here (for more detailed information including the source code itself see VoteBox (2009d).

VoteBox is described as a tamper-evident, verifiable electronic voting system created by researchers in the Computer Security Lab at Rice University (VoteBox 2009d). Although VoteBox is a prototype of an electronic voting system and may not be fully representative of all e-voting systems, which can contain a plethora of different functionalities and structures, it is a useful case study due to both its open source nature and the careful design and structure using best practice from computer science. Nonetheless, in common with all such computer systems, VoteBox assumes a boundary condition that means that the system assumes the right to vote by the actor that enters the voting booth. In the current implementation no identification checking is performed by the system itself – this remains external to the client voting system which merely records the vote and tabulates the final result for the returning officer.

The way in which a voter acts is visualised by means of a process or flow-chart. These shows the constraints on the system, its system boundaries and the general flow of information around the system. It is a common means of understanding complex data flows around a system within computer science and software engineering. A flowchart shows the system that would be experienced by the voter as they attempted to use the system. Flowcharts are very simple diagnostic and modelling structures that follow the logic of the program through a series of liner processes with decision gates, where a yes or no answer is expected, to guide the software to a certain resolution or output.

In the VoteBox system and documentation a lot of time is spent on a kind of voter ethnography where an idealised user, the 'voter', is given a great deal of attention. The analyses try to think about the way in which this user will operate the system, the kind of actions they might take and how they interface and use the technology. This involves a certain degree of ethnographic research by the designers, but also involves them creating a model of the human voter that runs through the entire system as an external constituent of it. The running code, then, is circumscribed by the 'running' voter, who is operating the voting machine, and must therefore by in 'sync' with the machine if it is to remain functional, correct and uncompromised by mistakes and errors. Much of the design, then, is spent on either the cryptography, which is the security layer for ensuring the vote is correctly stored and tamper-proof, but also on the user interface. For these designers, the running code, is always at the forefront of their minds as a voting system first, and then as a software system. As we discuss below, this idealised voter constantly seeps into the source code in a number of interesting ways, captured in the shorthand used in the commentary code and documentation. Although here we do not have the space to go into the interesting gender assumption that are being made, they demonstrate how programming commentary is important to analyse and take into account as part of the assemblage that makes up running software systems.

In sum for this e-voting system, a screen is displayed which gives to the voter a number of voting selections. Running code is here always code awaiting the vote of a user. The voter clicks their selections and the system passes to a review screen to show to the voter their selection. As the voter advances past the review screen to the final confirmation screen, 'VoteBox commits to the state of the ballot by encrypting and publishing it' (Votebox 2009d). Alternatively the vote might be 'challenged', compelling the system to reveal the contents of the same encrypted ballot.

This voting system consists of two main parts, a supervisor system that authorises the client systems, and the voting booths themselves which are likewise connected to the supervisor. In this example we will be concentrating on the voting booth software in particular and the way in which the programmers have, in the code, represented, often unconsciously, the nature of a typical 'voter/user' and the actions that are available to such an actor. The nature of the actor as understood by the system designer/programmer is interestingly revealed in two ways in the source code for the system:

```
/**
 * Allows the VoteBox inactive UI (what is shown when a user isn't voting)
 * to register for a label changed event, and update itself accordingly
 *
 * @param obs
 *              the observer
 */
public void registerForLabelChanged(Observer obs) {
    labelChangedEvent.addObserver(obs);
}
```

Figure 4.4 User represented in source code (VoteBox 2009c)

In the first code example given here, the actor utilising the voting machine is identified through the subject position of a 'user'. This is interesting on a number of different levels, but in the particularly discourse of computer programming one notes the key dichotomy created between the programmer and the user, with the user being by definition the less privileged subject position. The term user also carries a certain notion of action, most notably the idea of interactivity, that is that the user 'interacts' with the running software interface in particular circumscribed ways. Here though we also have the dichotomy of the user and the voter and often in programming systems there is a 'temptation of creating simplified and standardised models of the "citizen-user"' (Berry and Moss 2006: 31). One notes that in particular the simplified interface – actually more constrained in a number of ways than a simple pen and paper ballot as discussed below – already circumscribes the possible action of the voter/user.

```
currentDriver.getView().registerForOverrideCancelDeny(new Observer() {
            /**
             * Announce the deny message, and return to the page the voter was
             * previously on
             */
            public void update(Observable o, Object arg) {
                if (voting && override && !finishedVoting
                        && currentDriver != null) {
                    auditorium.announce(new OverrideCancelDenyEvent(mySerial,
                            nonce));
                    override = false;
                    currentDriver.getView().drawPage(pageBeforeOverride);
                } else
                    throw new RuntimeException(
                            "Received an override-cancel-deny event at the incorrect time");
            }
        });
```

Figure 4.5 'Voter' represented in the source code (VoteBox 2009c)

In this code sample, we see a slippage between subject positions of the 'voter' and the 'user'. In fact they appear to be being used interchangeably, although clearly they have clearly circumscribed actions associated with them. The 'voter' can cast a ballot within the terms of this system, although the above code fragment seems to indicate that when there is no vote being cast the actor is a 'user'. This would be interesting to observe as a running code system *in situ*, through, for example, an ethnography of the system in use.

```
/**
 * This is the event that happens when the voter requests
 * to challenge his vote.
 *
 * format: (challenge [nonce] [list-of-race-random pairs])
 *
 * @author sgm2
 *
 */
```

Figure 4.6 The male 'voter' represented in the source code (VoteBox 2009b)

A revealing moment by the programmer in this example, demonstrates that a particular gender bias is clearly shown when the programmer refers to the request of a 'voter' to challenge 'his' vote. In many ways this should not be surprising considering the fact that the vast majority of programmers are male, and this percentage is even larger in FLOSS culture. Nonetheless, we have revealed a number of interesting subject positions, the inactive 'user', and the male 'voter'.

```
/***
 * This is the strategy implementation for radio button voting. In radio button
 * voting, only one candidate can be selected in a race. Once a candidate is
 * selected, he can either be explicitly deselected, by being toggled, or he can
 * be implicitly deselected when the voter chooses another candidate in the
 * race. Selecting any candidate in any race implicitly deselects all other
 * candidates in the race.
 *
 */
public class RadioButton extends ACardStrategy {
```

Figure 4.7 The choice of the voter is technically constrained to only one candidate as represented in the source code (VoteBox 2009a)

Perhaps even more interesting, is the inability of the user-voter to cast a spoilt ballot, whether as an empty ballot, or a ballot that has more than one candidate selected. This is a technical condition built into the voting booth through the decision to use a particular digital object called a 'radio button' that prevent more than one selection being made. If this is attempted, then the previous choice is deselected. This decision will have consequences, for example in the ability of a voter to cast a protest vote or to exercise the right not to vote for any of the above. On the running software this could not be changed therefore forcing a choice on the voter, whose only option would be not to use the machine at all.

All of these small technical decisions act together to format the voting practice and provide a given set of processes and digital objects that are associated with them. They thus act to stabilise a particular instantiation of the voting process, rendering it more legitimate and material than other forms, additionally through the use of prescription the software, in effect disciplines the voter to act in particular ways (e.g. vote only once and select just one candidate) and circumscribes other forms of voting action (e.g. spoiling the ballot paper, leaving it blank, throwing it away, taking it home, etc.).

The question remains to what extent does the technical therefore effect the condition of possibility for being political? In the first instance, of course, there is the requirement for a certain technical or digital literacy in order to cast one's vote. Thus the political practice is now expressly reliant on a certain level of technical competence. Secondly, the voter must now rely on the correct inscription of their vote within the material substrates of the computer software and hardware and that these represent what Latour (2007) called 'immovable mobiles', that is that the vote remains stabilised throughout its passage from the booth to the data collection system (the supervisor in this case) and then on to when it is expressly counted. In the case of paper, there is always a paper trail, that is the vote can always be followed through the process by the human eye. In the case of software, the vote is encrypted and signed, such that this digital signature can indicate whether the vote has been changed or tampered with, however, once cast into the digital the only way to follow the vote is through its mediation through other software tools. By mediation I mean that the nature of mediation depends on the communicative function in social relations – that is, the possibility of communication. In this context, the enabling condition of mediation is the possibility of distance between the two points in the communicational process – here the voter and the returning officer. Here, it is the possibility of communication, rather than its actuality, that is crucial in understanding the communicational dimension of software, and, of course, the promise to deliver the message. One could say in this instance that all votes are mediated by software. Nevertheless this remains an issue of trust rather than merely a technical problem 'solved' by hard encryption or such like.

In this chapter, I have looked at some examples of how to analyse *running* code, namely through either a form of close (code in action) or distant reading (software in action). I therefore set out to use the interesting resources that the Internet has opened up for researchers in the example of both a cultural representation of code as performance, in effect running on the musicians that are instantiations of the code, together with an example from free software or FLOSS. Increasingly, due to the complexity of writing software large scale users are utilising the FLOSS databases of code in their work, and in doing so they present to researchers a unique possibility for looking at how code is used to implement political change. Above I looked at the VoteBox e-voting system code contained within the Google Code Repository, but there are huge quantities of software waiting to be unearthed and subjected to critical research.

The web itself, beyond its screenic representation presented by the browser offers a secret depth to the intrepid researcher that dares to use the menu function 'view source' or 'page source' on the View menu. Here the HTML code is revealed as much of the Internet is freely available to be examined and taken apart to see how code functions to construct certain forms of political subjectivity, action and digital rights.

This is not to say that examining running computer code can provide all the answers, and it is certainly not a replacement for existing methods for understanding political action and processes. However it is noticeable that little work is actually undertaken in relation to computer code and this could create fruitful new ways of thinking about politics when combined with existing analytical tools and the reality of running code today. This brief attempt to suggest ways in which running code might be analysed, particularly in conjunction with other methods, was deliberately written to avoid too much technical detail. However, running code, and the tools needed to analyse it, are in urgent need of humanities and social science approaches to both contextualise and deepen our understanding.

To look into this further and to broaden and deepen the question of code, I now turn explicitly to the questions raised through a phenomenological understanding of the computational, through a discussion of the work of Martin Heidegger, through a phenomenology of computation.

5
Towards a Phenomenology of Computation

Having suggested how the materiality of code might be subjected to critical analysis, I now want to focus on the experience of 'forgetting' technology. This raises the question of whether the experience of 'backgrounded' computational technology is as complete as we might think. Indeed, I want to explore the idea that technology is actually only ever partially forgotten or 'withdrawn', forcing us into a rather strange experience of reliance, but never complete finesse or virtuosity with the technology. Indeed, this forgetting, or 'being that goes missing', is for Heidegger 'the very condition of appearance (vanishing) of worldhood (Stiegler 1998: 244). Whilst I will go on to argue that there is something specific about the relationship that is set up between our use of digital devices and our experience of the world, I want to be clear that this is not merely to argue for a vulgar technological determinism. Such an approach was criticised by Raymond Williams (2003) who argued that,

> We have to think of determination not as a single force, or a single abstraction of forces, but as a process in which real determining factors – the distribution of power or of capital, social and physical inheritance, relations of scale and size between groups – set limits and exert pressures, but neither wholly control nor wholly predict the outcome of complex activity within or at these limits, and under or against these pressures (Williams 2003: 133).

Taking this into account, I want to develop the argument that we should not underestimate the ability of technology to act not only as a force, but also as a 'platform'. This is the way in which the loose coupling of technologies can be combined, or made concrete (Simondon 1980), such that the technologies, or constellation of technologies act as an environment that we hardly pause to think about. This is what Bertrand Gille calls the technical system, which 'designates in the first instance a whole play of stable interdependencies at a given time or epoch' (Stiegler 1998: 26). Gille explains,

> a technical system constitutes a temporal unity. It is a stabilization of technical evolution around a point of equilibrium concretized by a particular technology (Stiegler 1998: 31, emphasis removed).

For example, think of the way we use our mobile phones to manage our friendships through extensive database lists of numbers and addresses. Whilst we have the phone at hand, we can easily find where our friend lives to send a letter, or call them to have a chat. But should we lose the phone then we have lost not just the list of numbers, but also the practised habits of how we used to find information about our friends. Certainly this is the experience when one is then forced to upgrade to a new mobile phone, often complete with new software installed and frustrating new methods of interacting with it. This is a common experience with everyday digital technology to the extent that the constant revolution in interfaces is something that we have learned to accept, even if it is extremely frustrating, as we want the latest mobile phone, with all the perceived advantages of the latest technology. Further, when leaving Facebook due to the closed nature of the technology it is very difficult to extract your contacts, in effect meaning that Facebook attempts to hold onto your friends in order to hold onto you. Code is therefore used as a prescriptive technology.

I want to keep in mind that many previous thinkers have been overtly critical about 'new' technologies and their perceived effects on the minds and habits of human beings. For example, Plato, in *Phaedrus*, wrote that Socrates denounced the use of reading and writing, because those that use writing 'will introduce forgetfulness into the soul of those who learn it: they will not practice using their memory because they will put their trust in writing' or 'they will imagine that they have come to know much while for the most part they will know nothing' (Cooper 1997: 552). For Hieronimo Squarciafico, an Italian Humanist writing in 1477, 'printing had fallen into the hands of unlettered men, who corrupted almost everything', and he argued that an 'abundance of books makes men less studious' (Carr 2008). Similar arguments are being made today with regard to the deskilling of the mind that is purported to be the result of search engines, social media, and mobile technologies. Indeed, Nicholas Carr goes as far as to imagine that Google is responsible for making us both shallow and stupid (Carr 2008). We should therefore remain attentive to the temptation to see technology as leading to a decline in our abilities, indeed, in this chapter I specifically want to raise the question not of decline, but of a transformation in our being-in-the-world, even the possibility of a revolutionary one.

To look at the specific instance of computation devices, namely software-enabled technologies, I want to make a particular philosophical exploration of the way in which we *experience* digital technology. This is a method called *phenomenology*, and as such is an approach that keeps in mind both the whole and the parts, and that is continually reminding us of the importance of social contexts and references (i.e. the referential totality or the combined meaning of things). To undertake this phenomenology is to look at the way in which technology is already embedded in particular circumstances, and the constraints and opportunities that are locally available. For Wilfred Sellars (1962), the aim of philosophy is to understand things in the broadest possible sense, that is, to 'know one's way around' with respect to things in the world. Sellar's calls this web of reasons, justification, and intentions that enables us to negotiate the world a 'space of reasons'. So here we need to explore the way in which we can both know our way around technologies, but also the way in which technologies can shape what it is possible or us to know in the first place.

In this chapter, then, I want to understand in the broadest possible sense how to know one's way around *computationally* with respect to things in the world. This is a form of 'knowing *how*' as opposed to a 'knowing *that*', where one knows *how* to make a mobile telephone call, in distinction to knowing *that* the call is being transmitted via radio waves from the phone to a base-station. This 'knowing how to do something' in many ways increasingly presupposes access to a body of knowing-*that*, that is also knowledge of computational ways of doing things. In this case, one must have an embodied set of practices that frame and make available necessary knowing-that, towards which one is able to computationally know one's way around. This computational knowing-how is also tightly bound up with a class of objects I call technical devices, which themselves are able to perform a certain kind of knowing-how with respect to the human world. Technical devices are delegated performative and normative capabilities which they prescribe back onto humans and non-humans.[1] That is, a person lives in the midst of technical beings[2] that have specific forms of agency, or as Zuboff (1988) states, 'technology… is not mute',

It not only imposes information (in the form of programmed instructions) but also produces information. It both accomplishes tasks and translates them into information. The action of a machine is entirely invested in its object, the product. Information technology... introduces an additional dimension of reflexivity: it makes its contribution to the product, but it also reflects back on its activities and on the system of activities to which it is related. Information technology not only produces action, but also produces a voice that symbolically renders events, objects, and processes so that they become visible, knowable, and shareable in a new way (Zuboff 1988: 9).

For example, knowing-how to browse the world wide web, or knowing-how to use a satellite navigation system in a car calls for the user to think computationally in order to transform the inner state of the device such that it performs the function that is required of it.[3] In some senses then, one might argue that the user becomes an object of the technology, as Foucault argues, 'how does one govern oneself by performing actions in which one is oneself the object of those actions, the domain in which they are applied, the instrument to which they have recourse, and the subject which acts?' (Foucault, quoted in Fuller 2003: 140). Technology 'abstracts thought from action. Absorption, immediacy, and organic responsiveness are superseded by distance, coolness, and remoteness' (Zuboff 1988: 75). For example, distance becomes an abstract category within the navigation system; one is involved with programming the interface in a specific manner to achieve a specific goal, such as arriving at a set location by the shortest route.

An iPod shields you from the chaotic and unpredictable acoustic environment at large and indulges you with your favorite music. Amazon.com spares you the walk past buildings that depress you and people you'd rather not encounter on your way to the bookstore where a clerk will give you limited and unreliable information about the book you're interested in... The GPS device in your car makes it unnecessary for you to consult a map, stop at a gas station, count miles, or look out for signs and landmarks. The exterior world becomes irrelevant while computers keep the interior of your car pleasant and entertaining. Persons become tentative outlines when you meet them in reality and finely resolved images when you're back at your computer to Google what's of concern to you... The need to know is replaced by pieces of information that are summoned from nowhere and dissolve into nothing (Borgmann 2010).

But to fully interact with technical devices running code one is further encouraged to have some technical knowledge and an understanding of the collection of electronic resources that stand behind them. For example, our understanding of location itself is changed as we rely on the spatially constructions of satellites to make use of the devices; or even the knowledge that there is an isotropic world, within which there exists entities such as cars, satellites, computers, browsers, websites and so forth, that are mapped out in such a way as to have a computational orientation.[4] So, when one views the world computationally, one is already comported towards the world in a way that assumes it has already been mapped, classified, digitised. Space and place are constructed through computational devices which offer this world-view back through a plethora of computational mediators, such as mobile phones, car navigation systems, or handheld computers, for example,

> British researchers testing cognitive map formation in drivers found that those using GPS formed less detailed and accurate maps of their routes than those using paper maps. Similarly, a University of Tokyo study found that pedestrians using GPS-enabled cellphones had a harder time figuring out where they were and where they had come from... Cornell University human-computer interaction researcher Gilly Leshed argues that ... For the GPS users Leshed and her colleagues observed in an ethnographic study, the virtual world on the screens of their devices seemed to blur and sometimes take over from the real world that whizzed by outside. "Instead of experiencing physical locations, you end up with a more abstract representation of the world," she says (Hutchinson 2009).

The distantiation created by this collapse of distance instituted by computational technology alienates us from our local environment, as Heidegger states, 'all that with which modern techniques of communication stimulate, assail, and drive man – all that is already much closer to man today than his fields around his farmstead, closer that the sky over the earth, closer than the change from night to day...' (Heidegger 1966: 48).

The exemplar is perhaps the augmentation technologies that attempt to re-present reality back to the user via a picture of the world which is automatically overlaid with the result of computational geodata, tags and other content. This is reality experienced through the optic of a video camera, combined with an overlay of computer animation in real-time. Also known as 'air-tagging', 'spatial computing', 'optical internet', 'mixed reality', 'physical gaming', 'synthetic environments', and 'sit-sims'; the multiplicity of terms perhaps indicates the immaturity of the field, even though the term 'augmented reality' dates from around 1992. A good example is given by researchers in the INVENTIO-project, at the University of Oslo, who have created real-time situated simulations (Sitsims) of the cremation of Julius Caesar after the Ides of March as an example of the potential of the technology (Gliestoel 2010).[5] In other words, computational processes can extend and transform the lifeworld and 'also be used to craft possibilities that aren't simplified models of phenomena from our everyday world' (Waldrip-Fruin 2009: 4).[6]

Using Heidegger's notion of circumspection (*Umsicht* – 'looking-about'), which is a way of experiencing the world as an active being within a world full of meaning, I want to take seriously the idea that we can have an attitude towards the world, or better a 'way-of-being' which has a computational disposition of *circumspection*. In this case it is the computational aspect of the experience that I would like the pay particular focus on.[7]

This computational aspect is connected to an understanding of a world through a referential totality made up of technologies and information retrieval systems that make available to us an information-centric *familiarity* as part of our background experience. For Heidegger, this familiarity is a fundamental experience of the world as we do 'not normally experience ourselves as subjects standing over against and object, but rather as at home in a world we already understand' (Blattner 2006: 12). That is, we are not located in a system of objects, rather we live in a world, and to live in a world is to know one's way around it (Blattner 2006: 43). In the case we are discussing here, the contemporary milieu is suffused with technical devices with which we have to develop a familiarity if we are to be at home in the world. These raise important questions, for as Marx explains:

> Technology reveals the active relation of man to nature, the direct process of the production of his life, and thereby it also lays bare the process of production of the social relations of his life, and of the mental conceptions that flow from these conceptions (Marx 2004: 493, footnote 4).

Computation reveals a particularly rich set of active relations, between human and non-human actors, both collective and individual, as Fuller (2008) argues, 'the rise of software and computational and networked digital media in general has in many ways depended upon massive amounts of investment in institutions, training, and the support of certain kinds of actors (Fuller 2008: 6). Computation has moved from a small range of activities to a qualitative shift in the way in which we engage with knowledge and the world which highlights how important an understanding of the computational is today. The notion of computation, as universal computation, was itself,

> discovered by Alan Turing and described in his 1937 investigation of the limits of computability, "On Computable Numbers." A universal system can perform any computation that is theoretically possible to perform; such a system can do anything that any other formal system is capable of doing including emulating any other system (Fuller 2008: 269).

This idea of the universality of computability means that the range of applications and processes that are amenable to computation are startlingly wide, albeit later restricted with the notion of oracles, or uncomputable functions, see below (Hodges 2000; Turing 1939). Computability, for Turing, meant the *mechanization* of processes that could then be mathematically rendered and computed (Hodges 2000). Indeed, we live in a world of increasingly embedded computational devices which mechanise, stabilise and format the world through standardised formal processes. These technical devices also provide a form of distributed cognitive support for our access to, and understanding of, the world (both the social and natural world) by the nature of our becoming reliant on their computations. This raises the danger that we might 'suddenly and unaware…find ourselves so firmly shackled to these technical devices that we fall into bondage to them' (Heidegger 1966: 53–4). Of course, we have always used devices, mechanical or otherwise, to manage our existence, however, within the realm of digital computational devices we increasingly find symbolically sophisticated actors that are non-human. These devices are delegated particular behaviours and capabilities and become self-actualising in the sense of realising their potential by performing or prescribing complex algorithm-based action onto the world and onto us by acting to intervene in our everyday lives. As Carr (2010a) describes,

> I type the letter p into Google's search box, and a list of 10 suggested keywords, starting with *pandora* and concluding with *people magazine*, appears just beneath my cursor… Google is reading my mind—or trying to. Drawing on the terabytes of data it collects on

people's search queries, it predicts, with each letter I type, what I'm most likely to be looking for... It felt a little creepy, too. Every time Google presents me with search terms customized to what I'm typing, it reminds me that the company monitors my every move (Carr 2010a, original emphasis).

Of course, Google now has an 'instant' search, which even removes the requirement to press the Return key or click the search button, actively trying to guess what the user is trying to do, if not steer the direction of their thought. This demonstrates the very lack of withdrawal or semi-withdrawal of computational devices that I wish to explore in this chapter together with the phenomenological implications of this relationship. This is the phenomena of 'unreadiness-to-hand' which forces us to re-focus on the equipment, because it frustrates any activity temporarily (Blattner 2006: 58), that is that the situation requires deliberate attention. In the case of Google Instant, one would think that this might make the search process easier or more intuitive, but in fact the situation is quite the reverse, precluding the chance for the user to think about what it is they wish to search for. Conspicuousness, then, 'presents the available equipment as in a certain unavailableness' (Heidegger 1978: 102–3), so that as Dreyfus (2001a: 71) explains, we are momentarily startled, and then shift to a new way of coping, but which, if help is given quickly or the situation is resolved, then 'transparent circumspective behaviour can be so quickly and easily restored that no new stance on the part of Dasein is required' (Dreyfus 2001a: 72).[8] As Heidegger puts it, it requires 'a more precise kind of circumspection, such as "inspecting", checking up on what has been attained, [etc.]' (Dreyfus 2001a: 70). This is certainly the case with Google Instant, which with every keystroke constantly updates the screen, requiring more effort to check what has been typed and what is being shown.

In the broadest possible sense, how does one know one's way around *computationally* with respect to things in the world. First then, I want to examine the computational image as a particular, historically located, way of being through a phenomenology of computation (an ontotheology). Secondly, I want to explore the idea that we might have a mode of being within a computational image that is mediated through the action of computer code (whether real or postulated) that results in a distributed form of cognition and the form of the mediation provided by computation. Here I am drawing on the notion that is inherent within a computational view of the world that computational objects are equipment, but equipment as is a specific type of entity that does not withdraw. How, then, does one know one's way around the computational image?

More specifically, I want to consider what it means to negotiate a material world through the actions of a mediator, that is, through the agency of code, which can perform functions and actions which were previously within the realm of human action – the vicarious transformation of the entities within the world.[9] Finally, I want to draw out some of the political and philosophical implications of the emergence of such a way of being, and develop the notion of human agency as a distributed capability that goes beyond the somatic resources of an individual. Instead, I want to treat this agency as a variable outcome of a complex process of computation and transformation entangling both human and non-human actors (see Hutchins 1996). The critical question throughout is whether 'computation' is a concept seemingly proper to knowing-that has been projected onto knowing-how/Dasein and therein collapses the distinction between knowing-how and knowing-that hence inducing the substitution of knowing-that for Dasein.

Phenomenology and computation

In this section, I want to think through computation using Heidegger's existential phenomenology which aims to understanding different 'ways of being' through ontological categories of objects (beings), equipment (*das zeug*) and human-beings[10] (*dasein*); together with Sellars (1962) notions of the manifest and the scientific image[11] which are 'frames' for conceptualising phenomenal experience. For example, Heidegger considers the knowing-that to be a 'towards-which' that he calls *vorhandenheit*, or present-at-hand, which tends towards a scientific-rational perspective on understanding. This identifies substances and predicates and their formal relations to describe the universe; and the knowing-how to be a 'towards-which' he called *zuhadenheit*, or ready-to-hand, which is a comportment toward a special class of entities called equipment (*das zeug*) such that as a human-being you are responding to their affordances through finding your way around in the world. However, this knowing-that itself presupposes a world of knowing-how, towards which one acts in order to enable the construction of a body of knowledge of facts in the first place. Thus, for Heidegger, in agreement with Sellars, one 'image' is not prior or foundational to the other, rather they are co-constructive in the sense of mutually reliant on each other through a shared 'background' referential totality. Heidegger describes this background as:

> What is first of all "given"... is the "for writing", the "for going in and out", the "for illuminating", the "for sitting". That is, writing, going in and out, sitting, and the like are what we are a priori involved with. What we know when we "know our way around" and what we learn are these "for whats" (Heidegger 2010, *gesamtausgabe* band 21).

I want to suggest that what is happening in the 'digital age' is that we increasingly find a computational dimension inserted into the 'given'. Or better, that the ontology of the computational is increasingly hegemonic in forming the background presupposition for our understanding the world. For example, 'for writing' increasingly becomes 'for the processing of words towards-which a final document is output', or 'going in and out' becomes 'he exit or entrance towards-which one is involved in a process of attending to or withdrawing from a particular process'. Life experiences, then, become processual chains that are recorded and logged through streams of information stored in databanks. Experience is further linked to this through a minimal, decentred and fragmentary subjectivity which is unified through the cognitive support provided by computational devices which reconcile a 'complete' human being. I am not claiming that all aspects of experience will inevitably become computational, rather that our referential totality represented by the entities that surround us are increasingly actors enabled and pervaded with computational techniques which take on the referential model in a 'tertiary' or cultural form of memory (Stiegler 1998). This can be understood as a just-in-time memory provided by technical devices and structured by computational databases and processes. These technical devices have embedded within them a knowing-that which has been formalised and stored for the purposes of further computation together with methods which structure their agency. We might say that these devices call to us to have a particular computationally structured relationship with them. For Sellars and Heidegger, the body of knowing-*that*, the findings of which for devices becomes a dataset, is the domain of the special disciplines which provide a dataset for computational devices, these disciplines:

> ...know their way around in their subject matters, and each learns to do so in the process of discovering truths about its own subject-matter... the specialist must have a sense of how not only [their] subject matter,

but also the methods and principles of [their] thinking about it fit into the intellectual landscape (Sellars 1962: 35).

That is, that the methods and data are constructed to format the world in particular instrumental ways. For Heidegger, every discipline with a discrete subject matter is a 'positive science', and rests on an ontological 'posit' (a regional ontology) which is a presupposition about that the class of entities it studies *is* (Thomson 2003: 515). The computer scientist, for example, not only knows about computational methods and processes, but also what it is to think *computationally*. To distinguish between entities that are computational or may be represented or modelled computationally from the ones which are non-computational, computer scientists rely on an ontological understanding of what makes an entity computational, a sense of what Heidegger might have called the *computationality of the computational*. But also, in dealing with computational questions, for example, the computer scientist must face and answer questions which are not themselves, in a primary sense, computational questions, but deals with them to answer specifically computational questions which relate to the ability to select, store, process, and produce data and signals. One could therefore say that the for-what of computation is algorithmic transformation which connects to computer scientists own thinking in terms of its objectives, criteria and problems and how algorithmic methods are delegated into the wider culture.[12] In sum, *computer scientists attempt to transform the present-at-hand into the ready-to-hand through the application of computation.*

For Sellars, this process is undertaken through the synthesis of two pictures of great complexity, which bought together purport to be a complete picture of being-in-the-world. Sellars refers to these perspectives as the *manifest* and the *scientific* images of being-in-the-world (which we should understand in terms of Max Weber's notion of ideal-types). Thus we are faced with two conceptions, equally non-arbitrary, of being-in-the-world. Sellars argues, however, that we must try to understand how they are bought together into a single coherent experience, which he calls the stereoscopic. What Sellars is trying to draw our attention towards is the contradiction within the two images, whereby the manifest image presents a world of flow, continuous and entangled experiences, and the scientific image postulates a world of discrete elements, particles and objects. It is useful to think of Sellars's as offering a metaphysic which attempts to reconcile two images of the world and here I want to think through Sellar's conceptual schema as being a way of bringing to the fore the ontotheology of computationalism, that is, the historical specificity of a particular way of being-in-the-world for human beings. This attempt to reconcile the two images in computer science is linked to a notion of massive computational power in order to reassemble the shards of experience that technical devices capture into a continuous and seamless human experience. In effect, computation aims to perform this task by fooling our senses, assembling the present-at-hand objects together at a speed that exceeds our ability to perceive the disjunctures.

Here, it is useful to link Sellars' notion of the manifest image to the Heideggerian notion of being – or Dasein as the being that takes a stand on its own being and interacts meaningfully with equipment – and the scientific image to the notion of beings – as the present-at-hand of entities in the universe. Present-at-hand is experiencing 'use-objects as neutral, value-free entities with value added on [that] requires an artificial stance towards them, "a bare perceptual cognition", a "holding back from manipulation"' (Blattner 2006: 51) rather than understanding the ready-to-hand of equipment in terms of the role it plays in our acting in the world.

> The peculiarity of what is proximately ready-to-hand is that, in its readiness-to-hand, it must, as it were, withdraw in order to be ready-to-hand quite authentically. That with which our everyday dealings proximately dwell is not the tools themselves. On the contrary, that with which we concern ourselves primarily is the work – that which is to be produced at the time... (Heidegger 1978: 99).

The example given by Heidegger is that of the hammer, which we use as dasein in order to transform the world through a set of practices linked to a referential totality. In other words, the hammer is *for* something, it has meaning within a larger framework in which 'hammering' is understood as a practice related to a set of competences and knowledges, such as carpentry, and for this reason the hammer, as a single entity being used in the practices of hammering, withdraws from the foreground of experience. In contrast, technical devices cannot fully withdraw due to the internal modalities and instabilities of computational structures – the truth is that computational devices are brittle, unpredictable, and unstable (Weiner 1994: 4). This partial withdrawal, or unreadiness-to-hand, then, is what Blattner (2006: 58) calls a 'deficient mode' of readiness-to-hand, rather than being present-at-hand.

The manifest world is the world in which humans, or dasein, 'came to be aware of [themselves] as [being]-in the world', in other words, where humans encountered themselves as human (Sellars 1962: 38). This, in a certain historical sense, points to a notion of Special Creation, which argues that humans could not know themselves until they became human, and points to a fundamental discontinuity in the notion of a *manifest* image which is by itself irreducible (this certainly follows Heidegger's notion of a technological ontotheology, a kind of incommensurable Kuhnian moment). However, for Sellars, regardless of its historical emergence from what he calls the 'original' image, the manifest image is a refinement or sophistication of both empirical and categorical dimensions, whereby, it is

> the sort of refinement which operates within the broad framework of the [manifest] image and which, by approaching the world in terms of something like the canons of inductive inference defined by John Stuart Mill, supplemented by canons of statistical inference, adds to and subtracts from the contents of the world as experienced in terms of this framework and from the correlations which are believed to obtain between them (Sellars 1962: 40).

Therefore, the manifest image makes use of a scientific method which Sellar's calls 'correlational induction', but does not involve the postulation of imperceptible entities to explain the behaviour of perceptible entities. This clarifies that the manifest image of being-in-the-world is not a pre-scientific one, rather the manifest image is one of the poles of philosophical reflection towards which both speculative philosophy and systems and quasi-systems thinking have in common (Sellars 1962: 41). Fundamentally, the manifest image is a refinement of the original image construed as the progressive pruning of categories pertaining to the concept of personhood and their relation to other person's and groups of entities. Correspondingly, the emergence of modernity is the successive reclassification of entities as non-persons (e.g. trees, stars, planets, etc.) leaving only a human remainder.

The computational image

I would like to use Sellars' (1962) notion of 'image' to think through the comportment towards, and implications of, a computational set of categories and empirical knowledge which I will extend, calling this third form the *computational* image. Where Heidegger contrasts universe and world, and for Sellars this indicates the *scientific* and the *manifest* image, here I want to think through the possibility of a third image, that of the closed 'world' of the computer, the computational image. I want to understand how one know one's way around with respect to things in a computational image, and conversely, the computational way of making sense of the world and how it gives expression to that sensibility. Crucially, this involves a vicarious relation between computational entities through transformation and translations (understood as minimal transformations) which never directly encounter the 'autonomous reality of their components' (Harman 2009: 141). By vicarious, I mean acting or done for another, that is a mediation on behalf of another entity.[13] Harman (2009) argues that to operate vicariously 'means that forms do not touch one another directly, but somehow melt, fuse, and decompress in a shared common space from which all are partly absent' (Harman 2009: 142). Although Harman refers to a form of speculative realism, or object-oriented philosophy, that wishes to speculate on the same world as the sciences, I will restrict this notion to the computational image which shares features of both the manifest and scientific image in that the computational image contains both the discrete (i.e. *scientific*) and the continuous (i.e. *manifest*) dimensions or in some senses can form a bridge or interface between them. Here, I want to connect the computational image to Heidegger's notion of equipment, but crucially, I want to argue that what is exceptional about the computational device is that unlike other equipment which is experienced as ready-to-hand, computational devices do not withdraw, rather they are experienced as radically unready-to-hand.

So, how would one negotiate or cope with a world which is populated with equipment that is calling us through their affordances in particular contexts, think, for example, of mobile phones and iPods? These devices are prescribed with the facility to shape their environments in limited ways and to present a stable meaningful world, a towards-which we can give meaning to. The problem immediately arises for dasein that the physicality of the equipment is no longer familiar to us, it no longer shines. Rather, it acts as a carrier within which software is located and which as a plastic and black-boxed technology is both radically mutable and frustratingly fixed in form and function. Let us consider the iPod, a device that has been perfected in the form of the iPod Touch, iPhone and iPad, in which there is only one home button and a touch-screen interface which is context sensitive and infinitely reconfigurable. Here 'coping' or dealing with the device consists of being led through the narrative of the interface over which only limited control is available and therefore affordances may be promised but not delivered. An example of which is the fact that increasingly computational devices are not switched off – rather the screen is dimmed to give the impression to the user that the iPad is inactive. In reality, the device is merely waiting for the next interaction, which does not necessarily have to be with a human actor, for example it might continue to check for email, count the seconds for the clock, or update your location to a central computer server.

To compound the problem, the interface itself is liable to be upgraded, changed, or jailbroken (a term used to indicate that the user has escaped the boundary designated by the manufacturer). Thus to find one's way around remains a challenge where one is faced with the constant instability and unreadiness-to-hand (*unzuhandenheit*) of the computational interface. This unreadiness-to-hand, Heidegger argues, is a kind of partial present-at-hand (i.e. scientific image) which forces dasein to stop coping and instead sense a contextual slowing-down which Heidegger calls *conspicuousness*.[14] This is different to experiencing the world as manifest image, that is as a continuity of flow. Instead, it is a looking at things that appear to have come to a temporary fragmentary standstill, rather like when trying to learn a new skill when one must continually attend to what one is doing, such as when learning to ride a bike. Combined with the contemporary overly informatised environment, the deluge of information calling for attention may be overwhelming (see *New York Times* 2010). Psychologists call the requirement to move between different focal tasks a 'switch cost':

> [E]ach time you switch away from a task and back again, you have to recall where you were in that task, what you were thinking about. If the tasks are complex, you may well forget some aspect of what you were thinking about before you switched away, which may require you to revisit some aspect of the task you had already solved (for example, you may have to re-read the last paragraph you'd been reading). Deep thinking about a complex topic can become nearly impossible (Hopkins, quoted in Carr 2010b).

The iPad, like similar multitasked devices,[15] performs functions, which operate in a way that is both engrossing and frustrating, the device is at once too simple, presenting as it does a screen to the user which manifestly simplifies the underlying computational processes, and too complex, in that even with the shielding provided by the simplification the user often gets lost within tangled nested menus and options scattered across the device and which can be difficult to locate or even guess as to their function. Fuller (2003: 142) argues that the user therefore becomes an object of the technology, usefully pointing to the agentic nature of the technical device, but perhaps over-playing the extent to which the user is disabled by technology, although a partial distractedness is certainly often a result. One example of this is the inability of the device to present a unified interface within applications, therefore requiring the user to constantly move around in the application to try to find the controls or settings they are looking for.[16] Even more surprising is that 'scientists are discovering that even after the multitasking ends, fractured thinking and lack of focus persist. In other words, this is also your brain *off* computers' (Richtel 2010, original emphasis). So when surfing the Internet on the browser, the informational overload is astounding – people now consume up to 12 hours of media a day on average, many hours with multiple media simultaneously (e.g. TV and Internet). That compares with only five hours in 1960. Correspondingly, computer users now visit an average of 40 Web sites a day, according to research by RescueTime (Richtel 2010). Indeed,

> the Internet has a hundred ways of distracting us from our onscreen reading. Most email applications check automatically for new messages every five or 10 minutes, and people routinely click the Check for New Mail button even more frequently. Office workers often glance at their inbox 30 to 40 times an hour. Since each glance breaks our concentration and burdens our working memory, the cognitive penalty can be severe (Carr 2010c).

This places dasein in a relationship of towards-which that maximises the experience of conspicuousness perceived as a constant series of pauses, breaks, and interruption. One might reflect on a similar experience of the *obtrusiveness* of email, which, as anyone who has an email client on their work computer will be familiar with. As the disconcerting ease with which the visual or aural notifications continually break their flow, that is to move the user from a state of ready-to-hand, writing or using the computer to perform a task, to that of present-at-hand, which makes the entire computer apparent and available to inspection. Indeed, a study at the University of California, found that 'people interrupted by e-mail reported significantly increased stress' compared with those left to focus their attention on the text (Richtel 2010). Both conspicuousness and obtrusiveness, I want to argue, create a fragmentary and distracted flow of consciousness which, following Lyotard (1999: 5), I want to call a 'stream' and Deleuze and Guattari (2003) call the schizophrenic.[17]This is the *disjecta membra* of the human subject of the enlightenment and raises important questions about the *computational* subject in a contingent milieu which has an attendant devaluation of traditional and high culture. Here, I can only mark a connection I want to make with this 'fleeting-improvised' subjectivity and the computational image, that is, a subject that experiences conspicuousness as a continual state of exception and that is endlessly experiencing a promise of emancipation through the radical obsolescence of the socio-technological devices that surround it, discussed in more detail in the next chapter. Of course, this is the other side of the coin in that in the historical specificity of the computational way-of-being is offered the revolutionary potential of this recurring experience of infrastructural emancipation in a distributed notion of cognitive support through socio-technical devices, something that we will return to below.

The notion that equipment creates a state of conspicuousness for dasein gives the computational device its specificity and marks it out as radically different from other media, which are more comfortably ready-to-hand. For example, when one uses a hammer to strike a nail, for the carpenter, the hammer withdraws, providing the necessary conditions for such a tool are met (i.e. it is functioning correctly, not broken, not too heavy or too light, etc.). The carpenter can therefore use the equipment of the hammer without having a present-at-hand experience of the hammer which would get in the way of using it. Similarly, when watching television the audience forgets that it, as a medium, is there. In contrast a non-digital, analogue television is simple to use and presents a unified experience that withdraws so that the viewer can sit back and enjoy the show. When the television becomes digital, however, it is loaded with functionality, software, interfaces, menus and multiple options, such as the infamous red option button. The television is now a complex piece of machinery that needs constant care, careful management, and quite simply is capable of both crashing or corrupting whilst viewing but also interrupting the viewing experience (whether through digital techniques on the part of the broadcasters or locally with alarms, picture-in-picture, or other paraphernalia).[18]

For a computational device any withdrawal is partial, as it requires constant attention to keep it functioning and 'right' for the task it is to assist with, that is, a computational device remains in a state of conspicuousness. Even for something as static as an eBook reader, which only presents non-changing text to the user, the evidence suggests that the devices create a distracting object that users find difficult to concentrate on, in contrast to a physical book (see Hayles 2007 for evidence of a computationally reinforced 'hyper' attention state; see also

Kurniawan and Zaphiris 2001 for the specific problems of screen-based reading).[19] Here, I would like to connect Stiegler's (2009) conception of the fundamentally technical nature of this trajectory as *disorientation* with the growing cognitive assemblage represented by technical devices that now populate the manifest world, that is, that 'humanity's history is that of technics as a process of exteriorization' in which technical evolution is dominated by the technical inscription of memory as a method of reflective objectification. For Stiegler, it is the ability to place memory outside the body in some material form that gave rise to the possibility of reflexive thought and is a key aspect of how we came to be human in the first place. Here, I want to make the connection with the way in which modern computational technology is enabling the exteriorization of cognition and reflexivity itself.

The computational is also often closely associated with formal logic, calculation, and a particular type of rationality such as command and control, that is cybernetics (see Dubray 2009). In terms of computability, computation is exact when 'given exact finite data as input, an exact computation returns exact finite data as output' (Tucker and Zucker 2007: 2). Here, we can think in the first instance of a command-control model of usage whereby the user command the software to perform a task and the software willingly complies. This is certainly how most people understand their use of computers, the user remains in control. However, the near instantaneous translation of the command into action hides the discrete processes by which the command is converted into functions, checked against the software's internal checks and balances, and finally executed as an action. Within the domain of the computational processes, in the interstices between the manifest image and the digital representation is the possibility for the monitoring of and if necessary the realignment of the commands of the user. It is here the software acts to reflect the users desire, but the space between execution and feedback, which is given via the user interface need not represent the actual result, which again points towards the uncertain affordances of the computational device. As Carr (2010) writes,

> Eric Schmidt, Google's chief executive, once remarked that he looked forward to the day when Google would be able to tell him "what [he] should be typing," which, if I'm interpreting the statement correctly, also means that Google would be telling him what he should be thinking. Such a service, Schmidt said, would be the product he's "always wanted to build." (Carr 2010).

Indeed, the interface may perpetuate a form of software-ideology by misleading the user into anticipating the result of what Deleuze and Guattari (2003: 87) calls an order-word and which is expected to result in a transformation within the software or data. This act, the transformation of something into something else is an instantaneous act or what Deleuze and Guattari (2003: 80) call an incorporeal transformation. As they explain:

> The incorporeal transformation is recognizable by its instantaneousness; its immediacy, by the simulataneity of the statement expressing the transformation and the effect the transformation produces (Deleuze and Guattari 2003:81).

With software, however, the incorporeal transformation requested may not have been carried out, but the user is convinced by the software display that it has done so. This is a vicarious relationship, that is a relationship whereby following the command (order-words), the user transacts with the code to execute the action. It is a relationship that is mediated, and hence it is a relationship that separates the human from the world. One could think of a fly-wheel that acts to translate action and symbolic manipulation across multiple levels of digital code and which at any level the coupling in the assemblage may be tighter or looser depending on a number of factors (e.g. bugs in the code, hacking, mistakes in programming, or deliberate restriction of action through prescriptive code that is embedded or delegated with a particular normative content). This again highlights the importance of avoiding a screen essentialism if we are to open the black box of computational devices.[20]

Vicarious transformations

Vicarious indicates that there are interesting implications relating to the mediated relationship that we have with this 'hidden' computational world that is revealed only through transformation and translation of its internal functioning into a form that is projected into our phenomenal experience. This reminds one again of the object-oriented approach of Harman (2009:168), who describes parts encrusted onto a surface which are sensually available whereas 'the parts of a real object are contained on the interior of that object, not plastered onto its outer crust. In both cases, however, there is a vicarious cause enabling the parts to link together' (Harman 2009:168). This is, of course, similar to the notion of the human computer interface which connects the manifest image of the user to the internal world of the computational. In other words, there is no *direct* contact with our phenomenal reality and that represented within the computational device except through the interfaces, computer code, and input devices that mediate it, such as a mouse and a windowing system. As Heim (1987) explains:

> The types of physical cues that naturally help a user make sense out of mechanical movements and mechanical connections are simply not available in the electronic element. There are far more clues to the underlying structural processes for the person riding a bicycle than there are for the person writing on a computer screen. Physical signs of the ongoing process, the way that responses of the person are integrated into the operation of the system, the source of occasion blunders and delays, all these are hidden beneath the surface of the activity of digital writing. No pulleys, springs, wheels, or levers are visible; no moving carriage returns indicate what the user's action is accomplishing and how that action is related to the end product... The writer has no choice but to remain on the surface of the system underpinning the symbols. (Heim 1987: 131–2)

One example will suffice; in order to create the illusion of an interface that is presented to an ordinary user of the computer, there has to be a model of the multiple layers required, together with rules pertaining to the interrelation of parts of the screen and the masking and visibility of the components. We know this implicitly by the way in which things on-screen appear to vanish behind other things. However, in the world of the computational device this is merely an illusion, a screenic metaphor. Within the digital domain, discrete 'spaces' are created internally within the memory structures, held as voltage levels on memory chips, which function to draw out of a plane of immanence specific structures and namespaces that are independent of each other. As Kittler explains, '[a]ll code operations, despite such metaphoric faculties as call or return, come down to absolutely local string manipulations, that is, I am afraid, to *signifiers of voltage differences*' (Kittler 1997: 150, original emphasis). Through a process of abstraction and layering within the technical operation of the computer software there is a digital 'universe' in which digital entities are created as having discrete spatial characteristics, both in terms of occupying specific three-dimensional physical memory locations (i.e. on the memory chips), but also abstracting upon this physical space, a model of space that may have multiple dimensions and even contain alternative 'physics'. Therefore, it might be useful to think that digital devices 'have an Oreo cookie-like structure with an analogue bottom, a frothy digital middle, and an analogue top' (Hayles 2004: 75). Digital entities can then be said to have a double articulation in that they are represented both spatially within our material universe, but also with the representational space created within the computational device – a digital universe. The computational device is, in some senses, a container of a universe (as a digital space) which is itself a container for the basic primordial structures which allow further complexification and abstraction towards a notion of world presented to the user. That is, that each universe may itself by the conditions of possibility for further levels of abstraction and therefore further universes, rather like a Russian doll each within the other.[21] Here are clues to the basis for the claims of exponents of digital philosophy and the Regime of Computation.

The screen can be understood as a window onto this world and the keyboard and mouse operate as equipment with which we might manipulate it. However, this manipulation is never direct, and indeed the multiple levels of mediation are themselves rings which encircle the world into which we project a form of intentionality in terms of being-in-the-machine. As Hayles (2004) explains:

> the signifier exists not as a durably inscribed flat mark but as a screenic image produced by layers of code precisely correlated through correspondence rules, from the electronic polarities that correlate with the bit stream to the bits that correlate with binary numbers, to the numbers that correlate with higher-level statements, such as commands, and so on. Even when electronic hypertexts simulate the appearance of durably inscribed marks, they are transitory images that need to be constantly refreshed by the scanning electron beam that forms an image on the screen to give the illusion of stable endurance through time (Hayles 2004: 74).

We never directly encounter the entities that are constructed within the context of the digital world that is presented to us, instead, when we issue a command or move the mouse, a set of discrete translations are performed moving through the layers of the computational device to perform an uncertain transformation, which in the final instance involves the movement of voltage levels around the material circuitry within the computational device.

What strikes one as interesting about this process, is that the primordial elements of the computational device, the circuitry, the voltages, the silicon and so forth, appear to have no bearing on the actions of the user, in as much as the user concentrates on the world presented through the screen. On the other hand, the rules of physics must be attended to, and the quantum states of electrons, the electrical requirements of the environment within the computational device, the extreme requirements of the processor to be cooled and heat to be expelled are all handled invisibly by a process of worlding by the computer. In some sense the computational device has been delegated the capacity to gather these components and actively assemble them and continually stabilise their functioning. The computational device is an unstable form of equipment that must continually gather and reinforce its equipmental qualities against a hostile world of breakdown. This is then repeated through numerous layers of software that serve to create inner unstable universes within which further abstraction takes place, *all the way down*. Crucially though, the agency of each universe is loosely independent and defined at its creation in computer code by a series of constraints which serve as a framework within which the new abstract layer must function. Each layer promises uncertain affordances to the latter, eventually culminating in the partial affordance offered to the user through a risky encounter with a vicarious transformation which here I argue is radically unreadiness-to-hand.

This loose coupling of the user and the computational technical device offers possibilities that may be thought of in terms of Heidegger's notion of *Gelassenheit*. For Heidegger, *Gelassenheit* is a particular type of relationship with technical devices that is a letting go, 'serenity, composure, release, a state of relaxation, in brief, a disposition that "lets be." Seen from the standpoint of the will, the thinker must say, only apparently in paradox, "I will non-willing"; for only "by way of this," only when we "wean ourselves from will," can we "release ourselves into the sought-for nature of the thinking that is not a willing"' (Arendt 1971). This points towards the possibility of a relationship with technology that is not built of the will to power by virtue of the impossibility of control in a system that exceeds the comprehension of a human subject, this will be explored in the next chapter.

In any case, I have argued that the computational image problematically mediates between the manifest and scientific image and may hold important clues as to the difficulty of connecting or reconciling these images through its equipmental form. Of course, with the increasing reliance by physical sciences on technical apparatus, and the mediation of everyday life, particularly through digital devices, the computational becomes increasingly salient. The interesting point is that at the nexus of use, digital devices are a peculiar fragmentary mediator between the particulate and the continuous. In some senses then we might start to speculate on the nature of the computational image as a form of cultural analog–digital/digital–analogue convertor that translates entities between the manifest and scientific images but does so in an uneven and fragmentary way. In the next chapter this is extended through an examination of how this fragmentation combined with the huge quantity of data represented within the real-time data streams creates a new form of computational identity.

By way of conclusion, I suggest that by thinking about computationality, in particular code and software, as unready-to-hand, helps us to understand the specific experience of our increasingly code-saturated environment. Linked to this is the notion of a distributed form of cognition (we might think of this as a database of code enabled cognitive support), which we can draw on, like Google Instant, but which remains unready-to-hand. That is, that it causes us to suffer switching costs, which; even if imperceptively, change our state of being in the world. In the next chapter I want to look at how these forms of streamed-cognition are structured and some of the implications for how we might experience the world.

6
Real-Time Streams

The growth of the Internet has been astonishing, both in terms of its breadth of geographic cover, but also the staggering number of digital objects that have been made to populate the various webpages, databases, and archives that run on the servers. This has traditionally been a rather static affair, however, there is evidence that we are beginning to see a change in the way in which we use the web, and also how the web uses us. This is known as the growth of the so-called 'real-time web' and represents the introduction of a technical system that operates in real-time in terms of multiple sources of data fed through millions of data streams into computers, mobiles, and technical devices more generally. Utilising Web 2.0 technologies, and the mobility of new technical devices and their locative functionality, they can provide useful data to the user on the move. Additionally, these devices are not mere 'consumers' of the data provided, they also generate data themselves, about their location, their status and their usage. Further, they provide data on data, sending this back to servers on private data stream channels to be aggregated and analysed. That is,

1. The web is transitioning from mere interactivity to a more dynamic, real-time web where read-write functions are heading towards balanced synchronicity. The real-time web... is the next logical step in the Internet's evolution.
2. The complete disaggregation of the web in parallel with the slow decline of the destination web.
3. More and more people are publishing more and more "social objects" and sharing them online. That data deluge is creating a new kind of search opportunity (Malik 2009).

The way we have traditionally thought about the Internet has been in terms of pages, but we are about to see this changing to the concept of 'streams'. In essence, the change represents a move from a notion of *information retrieval*, where a user would attend to a particular machine to extract data as and when it was required, to an *ecology of data streams* that forms an intensive information-rich computational environment. This notion of living within streams of data is predicated on the use of technical devices that allow us to manage and rely on the streaming feeds. Thus,

> Once again, the Internet is shifting before our eyes. Information is increasingly being distributed and presented in real-time streams instead of dedicated Web pages. The shift is palpable, even if it is only in its early stages... The stream is winding its way throughout the Web and organizing it by nowness (Schonfeld 2009).

The real-time stream is not just an empirical object; it also serves as a technological *imaginary*, and as such points the direction of travel for new computational devices and experiences. In the real-time stream, it is argued that the user will be constantly bombarded with data from a thousand different places, all in real-time, and that without the complementary technology to manage and comprehend the data she would drown in information overload. Importantly, the user is expected to desire the real-time stream, both to be in it, to follow it, and to participate in it, and where the user opts out, the technical devices are being developed to manage this too. Information management becomes an overriding concern in order to keep some form of relationship with the flow of data that doesn't halt the flow, but rather allows the user to step into and out of a number of different streams in an intuitive and natural way. This is because the web becomes,

> A stream. A real time, flowing, dynamic stream of information — that we as users and participants can dip in and out of and whether we participate in them or simply observe we are [...] a part of this flow. Stowe Boyd talks about this as the web as flow: "the first glimmers of a web that isn't about pages and browsers" (Borthwick 2009).

These streams are computationally real-time and it is this aspect that is important because they deliver liveness, or 'nowness' to the users and contributors. Many technologists argue that we are currently undergoing a transition from a 'slow web to a fast-moving stream... And as this happens we are shifting our attention from the past to the present, and our "now" is getting shorter' (Spivak 2009). Today, we live and work among a multitude of data streams of varying lengths, modulations, qualities, quantities and granularities. The new streams constitute a new kind of public, one that is ephemeral and constantly changing, but which modulates and represents a kind of reflexive aggregate of what we might think of as a stream-based publicness – which we might call *riparian-publicity*. Here, I use riparian to refer to the act of watching the flow of the stream go by. But as, Kierkegaard, writing about the rise of the mass media argued:

> The public is not a people, a generation, one's era, not a community, an association, nor these particular persons, for all these are only what they are by virtue of what is concrete. *Not a single one of those who belong to the public has an essential engagement with anything* (Kierkegaard, quoted in Dreyfus 2001b: 77, italics added).

Here too, the riparian user is strangely connected, yet simultaneously disconnected, to the data streams that are running past at speeds which are difficult to keep up with. To be a member of the riparian public one must develop the ability to recognise patterns, to discern narratives, and to aggregate the data flows. Or to use cognitive support technologies and software to do so. The riparian citizen is continually watching the flow of data, or delegating this 'watching' to a technical device or agent to do so on their behalf. It will require new computational abilities for them to make sense of their lives, to do their work, and to interact with both other people and the technologies that make up the datascape of the real-time web. These abilities have to be provided by new technical devices that give the user the ability may therefore to manage this new data-centric world. In a sense, one could think of the real-time streams as distributed *narratives* which, although fragmentary, are running across and through multiple media, in a similar way to that Salman Rushdie evocatively described in *Haroun and the sea of stories*:

> Haroun looked into the water and saw that it was made up of a thousand thousand thousand and one different currents, each one a different color, weaving in and out of one another like a liquid tapestry of breathtaking complexity; and [the Water Genie] explained that these were the Streams of Story, that each colored strand represented and contained a single tale. Different parts of the Ocean contained different sorts of stories, and as all the stories that had ever been told and many that were still in the process of being invented could be found here, the Ocean of the Streams of Story was in fact the

biggest library in the universe. And because the stories were held here in fluid form, they retained the ability to change, to become new versions of themselves, to join up with other stories and so become yet other stories; so that unlike a library of books, the Ocean of the Streams of Story was much more than a storeroom of yarns. It was not dead but alive (Salman Rushdie, *Haroun and the sea of stories*, quoted in Rumsey 2009).

Of course, the user becomes a source of data too, essentially a real-time stream themselves, feeding their own narrative data stream into the cloud, which is itself analysed, aggregated, and fed back to the user and other users as patterns of data. This real-time computational feedback mechanism will create many new possibilities for computational products and services able to leverage the masses of data in interesting and useful ways. Indeed, we might begin to connect these practices of computational intensification with a wider computational economy which is facilitated by technology, which Kittler (1997) calls the technical a priori. These technologies may provide a riparian habitus for the kinds of subjectivity that thrives within a fast moving data-centric environment, and through a process of concretization shape the possibility of thought and action available. As Hayles (1999) states:

> Modern humans are capable of more sophisticated cognition than cavemen not because moderns are smarter... but because they have constructed smarter environments in which to work (Hayles 1999: 289).

Here, computational technology becomes instrumental to the processes of investment that individuals make into their lives, whereby success and intelligence is expressly linked to a technological process that makes these individual computational 'streams' more productive. The stream is also linked to the creation of a complex temporality through an assemblage of computational processes, through, for example the storage and recall of time-series data, a 'global' market-place and cycles of investment, dividends and company reporting requirements. These create wider oscillations which provide an informatised environment that is constantly changing but yet provides predictive patterns from seemingly random distributions of data.

The question now arises as to the form of subjectivity that is both postulated and in a sense required for the *computational* subject. In this final chapter, I want to think through the question of the subject as a computational stream, that is both a recipient of real-time data streams as a consumer and user of data and information, but also what a stream-like consciousness might experience. After spending the majority of the book thinking through the question of code and software through the optic of the computational, I now want to turn to the question of the computational subject. To do this, I want to look at the work of Jean-François Lyotard, a French philosopher and literary theorist, especially his ideas expressed in *Postmodern Fables*. Here, Lyotard introduces 'fifteen notes on postmodern aestheticization' (Lyotard 1999: vii). In these essays, he attempts to analyze the workings of the capitalist market through culture. His method shifts to a new 'subterranean practice' in which he moves from a commitment to the future anterior or the 'what will have been', (i.e. through experimentation by proceeding in such a way that the methods only emerge through the 'playing of the game' or after the event (see Beer and Gane 2004)), to a radical politics, or aesthetics, of disruption, using the fable as an exploratory approach. As Gane (2003) explains:

> The fable plays with the boundaries between fiction and reality, and in the process disturbs the narrative structures that frame and legitimate knowledge. Lyotard consequently terms fables 'realist', because they recount 'the story that makes, unmakes, and remakes reality' (Lyotard 1999: 91, quoted in Gane 2003: 444).

The fable is a narrative means of presenting a fictional or 'elusive ought', and at the end of the chapter I would like to consider what the moral of a postmodern fable of 'being a good stream' might be, much as in *Aesop's Tales* one is left with a moral at the end of the story. As Lyotard explains,

> In the fable the energy of language is spent on imagining. Therefore, it really does fabricate a reality, that of the story which it is telling; but the cognitive and technical use of reality is left pending. It is exploited reflexively, that is to say, sent back to language so that it can link up with its subject... Leaving it unsettled is what distinguishes the poetic from the practical and pragmatic (Lyotard 1993: 242).

But for now it is important to understand Lyotard's fables as part of a project of political resistance, where the poetic or mythic offers a line-of-flight through fleeting or disruptive movements. I want to use this as a means to think about computational subjectivity, that is, subjectivity that is mediated through computer-based technologies, in other words, a 'stream'-like subject. The problems introduced when our informationalised lives become mediated through the real-time is nicely captured by Borthwick (2009) who reflects that,

> The activity streams that are emerging online are all these shards — these ambient shards of people's lives. How do we map these shards to form and retain a sense of history? Like [that] objects exist and ebb and flow with or without context. The burden to construct and make sense of all of this information flow is placed, today, mostly on people. In contrast to an authoritarian state eliminating history — today history is disappearing given a deluge of flow, a lack of tools to navigate and provide context about the past. The cacophony of the crowd erases the past and affirms the present. It started with search and now its accelerated with the 'now' web. I don't know where it leads but I almost want a remember button — like the like or favourite. Something that registers something as a memory — as a salient fact that I for one can draw out of the stream at a later time. Its strangely comforting to know everything is out there but with little sense of priority of ability to find it becomes like a mythical library — its there but we can't access it (Borthwick 2009).

This concept of the stream as a new form of computational subjectivity also represents a radical departure from the individualised calculative rationality of *homo economicus* and tends rather toward the manipulation of what Brian Massumi calls 'affective fact', that is through an attempt to mobilise and distribute the body's capacity to think, feel and understand (either through a self-disciplinary or institutional form). Thus logico-discursive reasoning is suspended and replaced with a 'primary assemblage that links together statements, images, and passions in the duration of the body' (Terranova 2007:133). A link is formed between affective and empirical facts that facilitates and mobilises the body as part of the processes of a datascape or mechanism directed towards computational processes as software avidities, for example, complex risk computation for financial trading, or ebay auctions that structure desire. Indeed, the stream's comportment towards 'technical' or computational temporality and the connection between time, speed and movement for the maximization of output/profit lends it towards a form of subjectivity suited to the financialised practices that are becoming increasingly common today. This notion of computationally supported subject was developed in the notion of the 'life-stream':

> A lifestream is a time-ordered stream of documents that functions as a diary of your electronic life; every document you create and every

document other people send you is stored in your lifestream. The tail of your stream contains documents from the past (starting with your electronic birth certificate). Moving away from the tail and toward the present, your stream contains more recent documents — papers in progress or new electronic mail; other documents (pictures, correspondence, bills, movies, voice mail, software) are stored in between. Moving beyond the present and into the future, the stream contains documents you will need: reminders, calendar items, to-do lists... You manage your lifestream through a small number of powerful operators that allow you to transparently store information, organize information on demand, filter and monitor incoming information, create reminders and calendar items in an integrated fashion, and "compress" large numbers of documents into overviews or executive summaries (Freeman and Gelernter 1996).

This is a life reminiscent of the Husserlian 'comet', that is strongly coupled to technology which facilitates the possibility of stream-like subjectivity in the first place. Memory, history, cognition and self-presentation are all managed through computational devices that manage the real-time streams that interact with and make possible the life streams described here. These make use of the processing improvements associated with technology, together with feedback, control and rational management, which are reminiscent of cybernetic theory and the focus on information, feedback, communication, and control (Beniger 1989). It is also argued that what we see are changes in the internal structure of the human mind and body to facilitate that productivity that previously took place in the factory (Hardt and Negri 2000). This is the restructuring of a post-human subjectivity that rides on the top of a network of computationally-based technical devices. This notion of a restructured subjectivity is nicely captured by Lucas (2010) when he describes the experience of dreaming about programming,

> This morning, floating through that state between sleep and consciousness in which you can become aware of your dreams as dreams immediately before waking, I realized that I was dreaming in code again... [D]reaming about your job is one thing; dreaming inside the logic of your work is quite another... But in the kind of dream that I have been having the very movement of my mind is transformed: it has become that of my job. It is as if the repetitive thought patterns and the particular logic I employ when going about my work are becoming hardwired; are becoming the default

logic that I use to think with. This is somewhat unnerving (Lucas 2010: 1).

This is the logic of computer code, where thinking in terms of computational processes, as processual streams, is the everyday experience of the programmer, and concordantly, is inscribed on the programmer's mind and body. The particular logic of multiple media interfaces can also produce a highly stimulated experience for the user, requiring constant interaction and multi-tasking. According to Richtel (2010),

> heavy multitasking might be leading to changes in a characteristic of the brain long thought immutable: that humans can process only a single stream of information at a time. Going back a half-century, tests had shown that the brain could barely process two streams, and could not simultaneously make decisions about them. But Mr. Ophir, a researcher at Stanford University, thought multitaskers might be rewiring themselves to handle the load... [however actually] they had trouble filtering out... irrelevant information (Richtel 2010).

These are interventions that are made possible through new media technologies, such as word-processors, project management software and intimate technologies like the iPhone, technologies that provide an environment in which thinking is both guided in a logical fashion, but also continually fragmented across the media interface. This can change the very act of writing itself, as Heim writes:

> You no longer formulate thoughts carefully before beginning to write. You think on screen. You edit more aggressively as you write, making changes without the penalty of retyping. Possible changes occur to you rapidly and frequently... The power at your fingertips tempts you to believe that faster is better, that ease means instant quality (Heim 1993: 5).

It is this constantly present form of subjectivity that is closely linked to the computational experience of technical devices described above. These, of course, are highly dependent upon the code that makes up the data processing component that enables the streams in the first place. By displacing certain activities into the technology enables rapid reflexive augmentation of the data that is in a constant feedback loop back to the user. This is the human being as a data stream in its own right, or as is more commonly termed, a user stream.

Being a good stream

Lyotard develops his notion of the 'stream' from his previous works, *The Postmodern Condition* (1984), and *The Inhuman* (1993), where he drew attention to the rapid pace of technological change and its potent possibilities to extend rationalisation and domination. In *Postmodern Fables* he is expressly interested in technology's ability to speed up the exchange of information to such an extent that critical thought itself might become suppressed under the quantity of information. For example, in the first essay called 'Marie goes to Japan', Lyotard tells the tale of Marie, a overworked academic who must travel the world in order to 'sell her culture' and in doing so, becomes a 'stream of cultural capital: a member of a new "cultural labour force" that is exploited by choice' (Gane 2000: 444). Lyotard explicitly links economic value and speed, indeed, as he explains in the note: 'capital is not *time is money*, but also *money is time*. The good stream is the one that gets there the quickest. An excellent one gets there almost right after it has left' (Lyotard 1999: 5). For the 'good little stream' of the fable, it is the ability to produce rapidly that is the key marker of success, indeed, the faster something is completed and thus increases the stream's flow, the more profitable, the more successful and the greater the level of productivity. As Lyotard remarks:

> The best thing is to anticipate its arrival, its 'realisation' before it gets there. That's money on credit. It's time stocked up, ready to spend, before real time. You gain time, you borrow it.(Lyotard 1999: 5).

This improvement in the 'efficiency' of the individual recalls Marx's distinction between absolute and relative surplus value and the importance to capitalism of improvements in both organizational structure and technological improvements to maximizing profit (Marx 2004: 429–38). So, for example in this case, writing academic papers and books, using technology in any spare moments of time, together with mobility and participation, are the key to understanding this intensive new world of cultural production. However, this production is in a sense cut off from a sense of history, what Bruce Sterling calls the 'atemporal' (2010). This is constantly generating new forms of cultural capital through networked activity in the radical present and whose success or failure is judged in reference to current continual output. This is a form of production that is built around a normative ideal of continual work, continual streams of discrete quantifiable products that can be distributed and which feed into other shared work. As the Invisible Committee (2009) noted,

> Ideally you are yourself a little business, your own boss, your own product. Whether one is working or not, it's a question of generating contacts, abilities, networking, in short "human capital" (The Invisible Committee 2009: 50–1).

But there is not just a relationship between the quantity of time spent on the project and the resultant success; rather, it is the compression of time, the raising of productivity and efficiency that is important. It is the reduction in total time between the inputs and outputs of a process that Lyotard is drawing attention to as, following Marx, 'moments are the elements of profit' (Marx 2004:352). In the computational, the moments are not measured in working days or hours, but rather in the 'technical time' of the computer, in milliseconds or microseconds. It is here, technologies are inserted into cultural production in order to speed-up the creation of culture and its circulation. This is related to what economists call Total Factor Productivity (TFP), that is, where technological advances have lead to a continual increase in productivity, rather than a reliance on increased capital and labor inputs. Lyotard explains, 'you have to buy a *word processor*. Unbelievable, the time you can gain with it' (Lyotard 1999: 5). This brings to mind the experience of Friedrich Nietzsche who in 1882 after having bought a typewriter to help him write due to his failing vision, found that 'our writing equipment takes part in the forming of our thoughts' (Kittler 1999: 201). Indeed, 'in 1874, eight years before he decide[d] to buy a typewriter, Nietzsche ask[ed] himself whether these are still men or simply thinking, writing, and computing machines' (Kittler 1987: 116).[1]

Materialising the stream

To be computable, the stream must be inscribed, written down, or recorded, and then it can be endlessly recombined, disseminated, processed and computed.[2] The recording includes the creation of collective notions of shared attributes and qualities, in many cases institutionally located and aggregated,[3] but also a computational narrative of the subject through the datascape specifically represented through the data points they collect through their lives, either privately as geodata, twitter feeds or such like, or publicly through health records, tax records or educational qualifications.[4] The consistencies of the computational stream are supported by aggregating systems for storing data and modes of knowledge, including material apparatuses of a technical, scientific and aesthetic nature (Guattari 1996: 116).

These link directly to some of the issues raised by the body of work that has come to be known as medium theory, including Hayles (2005), Kittler (1997) and McLuhan (2001), that tries to think through the question of *storage* through the invention of 'new materials and energies, new machines for the crystallizing time' (Guattari 1996: 117) – particularly relevant in regard to the processes of computational flows. Here, I am not thinking of the way in which material infrastructures directly condition or direct collective subjectivity, rather, the components essential 'for a given set-up to take consistency in space and time' (Guattari 1996: 117).[5] We might think about how the notion of self-interest is materialised through technical devices that construct this 'self-interest', for example, through the inscription of accounting notions of profit and loss, assets and liabilities, which of course increasingly take place either through computer code which is prescribed back upon us.

It is important to consider the question of storage with regard to the computational stream. It is also crucial that a link is made between the computational and storage, as computation requires both the processing code and the data to be inscribed somewhere. This requires a chain of signification as 'memory' to be generated which translates the stream of data into a symbolic order through code. This technical a priori is crucial to understanding what it is possible to record at all, and the medium that translates and stores the data that forms the 'memory' of the computational. Here, we can think of computation requiring a network of writing which creates computable numbers that are divided into discrete countable finite elements. In other words, computational data is artifactualised and stored within a material symbolisation. This computational network requires a material channel through which the media of computation are carried, but as Kittler (1997) notes, it is a characteristic of every material channel that beyond, and against, the information it carries, it produces noise and nonsense. We have the assemblage of a network which builds the material components into an alliance of actors and which is a referential totality for the meaning that is carried over it, and past its borders, policed by human and non-human actors, we have what Doel (2009) calls *excess* and Latour (2005) calls *plasma*.

Here, then, we see the movement or translation between the temporal generation of the discrete elements of the stream and the computational storage through what Kittler calls *time axis manipulation*. This is the storing of time as *space*, and allows the linear flow to be recorded and then reordered. The shifting from chronological time to the spatial representation means that things can be replayed and even reversed, this is the discretisation of the continuous flow of time. Without it, the complexity of financial markets would be impossible and the functions and methods applied to it, through for example the creation of new abstract classes of investment such as Credit Default Swaps (CDSs), Collateralised Debt Obligations (CDOs) and Asset Backed Securities (ABSs) would be extremely difficult, if not impossible, to create and trade.

This implicit datastream across all devices leads to an enormous amount of data being collected and held in corporate databanks and huge data centres. As Borthwick (2009) noted, bit.ly, an Url shortening site, had collected 200 gigabytes of click data by 2009, including: usage data, location data, and so forth, about the users of the site which the users would not be aware had been collected. This is the idea of a 'dataspace', richly endowed with content which is dereferentialised and equally accessible by being located within a database and which makes the presence of data seem addictive and overwhelming. As Borgman notes,

> The glamorous fog of cyberspace varies in thickness. It's denser when we sit in front of the computer than when we are face to face with a person. It's thinner for the driven and the ambitious than for the sullen and the addicted. But when it is thick, it's disorienting in a new and distinctive way. The problem is not that we can't find what we are looking for, but that we are not sure what to look for in the first place. Whatever we have summoned to appear before us is crowded by what else is ready to be called up. When everything is easily available, nothing is commandingly present (Borgman 2010).

This notion of the computational dataspace is explicitly linked to the construction of the stream-like subject and raises many important questions and challenges to the liberal humanist model of the individual. Most notably in their bounded rationality – here the information and processing to understanding is off-loaded to the machine – but also in the very idea of a central core of human individuality. It also returns us to the question of digital *bildung* and how we structure the kind of education necessary in a computationally real-time world. For example, although,

> most streams today are explicitly created by users, either by creating content, making a friend, saving a favorite etc. For every explicit action of a user, there are probably 100+ implicit datapoints from usage; whether that is a page visit, a scroll, a video/shopping abandon etc (danrua, comment in Borthwick 2009).

However, we must not lose sight of the materiality of these computational forms which is inscribed within a material substrate. This is a computationally generated digital world that is limited by certain material affordances in the use of technologies such as processor capacity (i.e. computers do not have an unlimited amount of time nor do they have infinite storage space). Computation creates a technical form of time through the conservation, accumulation and sedimentation of past stream data, what we might call its memories or its past, which is then rearticulated in light of unfolding new data computation. This is what Heidegger (1988: 260) called the technical measurement of time, the attempt to determine the undetermined through the recording of the past through an apparatus of inscription. Without preservation, there is no stream, as it would be a mere atomistic point in time. Without the storage and recall of data there is no computational possibility for the construction and action of a focal attention by the stream. Together, these devices form complex assemblages that entangle the user/stream into a particular memory-temporality which creates the conditions for particular kinds of agency.

Heidegger considered 'authentic' time to be time in relation to death, as finitude and mortality. In the time of the computational stream, however, time is found in the inauthentic time of measurement, the attempt to determine the 'undetermined' through technical devices. So, for example, in the case of the computational, there is only the abstract notion of time as reported through the continual ticks of the datastreams and the charts and visualisations that represent the time-series datascape to the viewer.

The notion of subjectivity that is embedded in the socio-technical networks of computational systems points towards a deathless existence, even as, paradoxically, the market continually relies on the anxiety represented through sickness and disease, poverty and old age, to activate and fuel desire. These technologies operate to create and sustain a market which introduces a time of indeterminacy and choice into the stream of flows even as they stimulate affective responses within the stream calling for forms of action. For financialised streams, for example, the ticks of financial data are linked to the body as the profit or loss of securities and entangled with desires, necessity and ontological security. For the user stream, it is a constant flow of everyday activity represented as a chaotic uncoordinated stream of events logged to a microblogging site.

These streams are undoubtedly creating huge storage issues for the companies that will later seek to mine this collection of streamed data. For them, the problems are manifested in the building of massive computational data centres in locations around the world. Trying to capture the ready-to-hand world of everyday life generates such a large flow of data that can easily overwhelm these systems, witness the intermittent downtime of services like Twitter, which are also forced to regulate the flow of data into their networks through API feeds. Connected to this storage medium are the processing practices that are applied to render the stream of computational data as a source of action. These allow the analysis and visualisation of computational patterns over time, and allow the discernment of trends and traces that are left as markers within the data. There is a growing and important literature on the issue of data visualisation in general (see Pryke 2006; Manovich 2008), and financial markets in particular (Beunza and Stark 2004; Beunza and Muniesa 2005; Knorr Cetina and Bruegger 2002), here, I can only note the importance of the visual mediation of this data and its highly aestheticised content but clearly with the amount of data available the skills of a visual rhetoric will become increasingly important to render the patterns in the data meaningful.

Using financialisation as an example of a type of computational subjectivity, we might link the movement of a calculative rationality to that of an affective distributed rationality, geared towards the consumption of a financialised range of goods and services. I mark and develop the notion of the stream in the section below through a discussion of the notion of financialisation and a tentative cartography of the subjectivity associated with it, which I connect to the 'degradation of the individuals capacity for *understanding* their own circumstances, and their ability to make any effective use of whatever *correct understandings* they might achieve' (Terranova 2007: 132, original emphasis) – here particularly through the dichotomy of pattern/randomness (and here I want to connect randomness to a notion of plenitude). I want to think about the way in which life itself becomes understood as a 'life-stream' through the application of memory systems designed to support a highly informatised and visualised computational economy. That is, I want to understand the stream as a 'propagation of organised functional properties across a set of malleable media' (Hutchins 1996: 312). Connected to this are notions of calculability and processing, which relate back to the creation of technical devices that facilitate the user's ability to make sense of the movements in markets, data, and culture and more particularly, to respond to changes in risk and uncertainty. Users treat their lives as one would a market portfolio, constantly editing the contents through buying and selling, creating new narratives through the inclusion or exclusion of certain types of product or data stream.

Financial streams

Financialisation is an analytical term used to describe the processes of finance capital, including the institutions, norms, practices and discourses that are connected with it. It is thus a useful means to unpack the way in which claims to an information society or knowledge economy are bound up with particular situated approaches to organising the economy, society and politics. Financialisation has implicit within it, certain ways of acting, certain ways of being and certain ways of seeing that are connected to a particular comportment to the world, one that is highly attenuated to notions of leverage, profit and loss and so forth. Moreover, financialisation implies that the rational actor of economic theory is transformed from the calculative rationality of the protestant work ethic to an actor that is guided not only by rational self-interest but also a propensity to understand and take highly-leveraged and complex risks. That is, to move beyond Weber's description of the religious basis of capitalism as 'exhort[ing] all Christians to gain all they can, and to save all they can; that is, in effect, to grow rich' (Wesley quoted in Weber 2002: 119, emphasis removed). Where Weber described monetary acquisition as saving linked to an ethical norm supplied by protestant faith – that is, an understanding as labour and saving as a calling – with financialisation we see quite the opposite with a move towards the use of debt financing to fund investment and consumption to the extent that its lack of ethical grounding arguably leads inexorably towards endogenous financial instability through Speculative and Ponzi modes of investing (Minsky 1992). In a different register, Belfrage (2008: 277) glosses financialisation as 'emerging out of conditions which force people to weigh up the market performance of their financial assets when making everyday decisions between saving and consuming'. Financialisation is, nonetheless, an essentially contested concept, and as Randy Martin (2002) explains:

> Financialisation, like those other recently minted conceptual coins postmodernism and globalization, gets stretched and pulled in myriad directions. Part of the complexity of these terms is that they stand simultaneously as subject and object of analysis—something to be explained and a way of making sense out of what is going on around us.

Here, I follow the work in the sociology of markets to understand financialisation as the uneven process of formation of a socio-technical network that is used to stabilise a certain kind of calculative cognitive-support, that mediates the self and the world through financial practices, categories, standards and tests (Callon 1998). More importantly, I want to link the processes of financialisation to the creation of rapidly changing data streams of financial information. Rather than restricting the notion to the purely discursive or economic, I want to tentatively explore the idea that financialisation itself is the establishment of valuation networks; that is, the construction of circuits of finance which render abstract financial objects commensurable and exchangeable, in which actors, both human and non-human, are enrolled. This is in distinction to cognitive psychology that sees the ability for actors to calculate as being either rendered within a form of mental calculation which cognitive anthropology, has shown to be far too demanding, and also in distinction to cultural approaches which see the calculative competence through social structures or cultural forms and which is unable to explain the shift from one form of calculative agency to another (Callon 1998: 4–5). Further, I want to challenge the notion of a linear process of financial transformation – 'financialisation', and instead highlight the way in which there is an assemblage of *'financial mediation'* itself marked by a series of tensions, counter-tendencies and modulations.

A financialised assemblage is connected together through the use of equipment or financial computational devices (what Deleuze would call *agencements*) whose aim is to maintain an anticipatory readiness about the world and an attenuated perception towards risk and reward which is mediated through technical affective interfaces (i.e. the computer user interface). In the first place, the computational is directly linked to quantitative statistical processing of massive amounts of time-series data and its visualisation or representation. Additionally, however, the affective dimension seems to me to be extremely important in understanding the way in which recent shifts in financial markets towards the democratisation of access have been intensified through the realignment of desire with the possibilities offered through monetary returns from finance capital – what Bloom (n.d) has called 'computational fantasies' – and it is a subject I'll return to below. But none of these practices of intensification could have been possible without information technology, which acts as a means of propagation but also a means of structuring perception – or better, of 'focusing' attention in the sense of an extended mind. Finance itself has a 'feel to it' which is generated via the computer interface or through the marketing and packaging that 'wraps' the underlying financial product.

This affective dimension to finance is also interactive, providing a model of action that situates the user (or investor) in a relation of continual interaction with their portfolio. The important point here is that you do not need to have a 'whole' human being who has intentionality and therefore makes rational decisions about the market, or has feelings and is responsible for their actions and so on. Rather, you can obtain a complete human being by composing it out of composite assemblages which is a provisional achievement, through the use of computer cognitive support (what Latour (2005) neatly calls 'plug-ins') and we might think of as software interfaces or technical devices. For example, these are the share-trading systems that initially pre-format the user as a generic market investor. But to be an active investor requires the use of particular techniques and strategies in the market supported through extra software interfaces that offer guidance on 'reading' the market (see for example the websites: The Motley Fool, or Interactive Investor). One example of this is that of Swedish pension reform, where individual pension investment is part of a process amenable to 'nudges' by technical devices that help guide the individual through up to 1000 different investment funds (Thaler and Sunstein 2009).

These can be understood as structuring templates that act as devices to give you the capacity to calculate, that is, cognitive abilities that do not have to reside in 'you' but can be distributed throughout the investment interface. It is important to note, however, that the extent of the 'nudge' that the system can provide can range from the libertarian paternalism of defaults and formatting advocated by Thaler and Sunstein (2009) to posthuman distributed aids to cognition, or even collective notions of cognition, as described by Hutchins (1996). An example is the portfolio manager software offered by a number of companies online, which purport to not only hold the investment portfolio, but rather to stimulate you to invest, trade and have a way of being-towards the market which is active (Interactive Investor is a notable web-based example). This can be achieved through email alerts set to certain time-series prices, automatic trading systems and constant feedback to the user via mobile technologies (see the Stocks.app application on the Apple iPhone, for a mobile example). Wherever the investor is, they are able to call up the portfolio and judge their asset worth as defined by the external forces of the financial markets but crucially simplified and visualised through the graphical capabilities of the mobile device (see tdameritrade, Etrade, iStockManager, m.scottrade.com, etc.). Sometimes, in a radical break with the notion of judgement being the seat of humanity and contra Weizenbaum (1984), the software can even judge the success of the investment strategy through a number of algorithmic heuristics, something the investor may not even have the calculative or cognitive ability to challenge.

In the case of financial markets, software has completely changed the nature of stock and commodity markets creating 24 hour market trading and enabling the creation of complex derivative products and services, often beyond the understanding of the traders themselves. For example, high frequency trading (HFT) is at the cutting edge for trading on financial markets, the basic idea of HFT is to use clever algorithms and super-fast computers to detect and exploit market movements. To avoid signalling their intentions to the market, institutional investors trade large orders in small blocks—often in lots of 100 to 500 shares – and within specified price ranges.

> High-frequency traders attempt to uncover how much an investor is willing to pay (or sell for) by sending out a stream of probing quotes that are swiftly cancelled until they elicit a response. The traders then buy or short the targeted stock ahead of the investor, offering it to them a fraction of a second later for a tiny profit (*The Economist* 2009).

These changes in the practices of stock market trading reflect the implementation of high technology networks and software, indeed,

> HFT is a type of algorithmic trading that uses high-end computers, low-latency networks, and cutting-edge analytics software to execute split-second trades. Unlike long-term investing, the strategy is to hold the position for extremely short periods of time, the idea being to make micro-profits from large volumes of trades. In the US, it is estimated that 70 percent of the trade volume is executed in the HFT arena (HTCWire 2010).

This technology came to public attention on 6 May 2010 when the Dow plunged nearly 1,000 points in just a few minutes, a 9.2 per cent drop, and christened the 'flash crash'. Half a trillion dollars worth of value was erased from the market and then returned again. Due to the work of software engineer Jeffrey Donovan, it became clear that HFT systems were shooting trades into the markets in order to create arbitrage opportunities. By analysing the millisecond data stream logs of the exchange and reverse-engineering the code, he was able to see the tell-tale signs of algorithmic trading in cycles of 380 quotes a second that led to 84,000 quotes for

300 stocks being made in under 20 seconds, which set off a complex chain of reactions in the market and the resultant slump (HTCWire 2010).

Financial companies are rolling out new experimental technologies continually to give them an edge in the market place; one example is the so-called 'Dark Pools' (also known as 'Dark Liquidity'). These off-market trade matching systems work on matching trades on crossing networks which give the trader the advantage of opaqueness in trading activities, such as when trying to sell large tranches of shares (Bogoslaw 2007). Dark pools are 'a private or alternative trading system that allows participants to transact without displaying quotes publicly. Orders are anonymously matched and not reported to any entity, even the regulators' (Shunmugam 2010). Additionally, technologies such as 'dark algorithms' give firms the ability to search multiple dark pools to find hidden liquidity.

Software that acts in this cognitive support capacity can therefore be said to become a condition of possibility for a device-dependent, co-constructed subjectivity. This Guattari (1996: 114) calls a 'processual' subjectivity that 'defines its own co-ordinates and is self-consistent' but remains 'inscribed in external referential coordinates guaranteeing that they are used extensively and that their meaning is precisely circumscribed' (Guattari 1996: 116). The subject, then, is circumscribed by the technologies which mediate its relationships with finance capital, such that the field of experience is constantly shifting to reflect financial data and the movement of time. Following Lyotard, we might declare that the subject becomes a computational 'stream', in this case a stream attenuated to the risk associated with finance capital mediated through financial software.

Of course, risk itself is a pivotal category in modern finance that is stabilised through the use of technology and discourse. Risk, for Langley (2008), is distinct from uncertainty, where uncertainty is understood as non-calculable future volatilities that are beyond prediction, and risk itself is a statistical and predictive calculation of the future. Langley explains:

> There is no such thing as risk in reality... risk is a way – or rather, a set of different ways – of ordering reality, of rendering it into a calculable form it is a way of representing events in a certain form so that they might be made governable in particular ways, with particular techniques and for particular goals' (Dean quoted in Langley 2008: 481).

This is, of course, the notion of risk developed by the economist Frank Knight in his 1921 book *Risk, Uncertainty and Profit*. When encoded into financialised software, risk is qualified, rendered and abstracted in a calculative space which interfaces to the investor through devices that seek to present the world through what Taleb (2007) calls Gaussian risk. This is risk that is presented without its limitations as a model made transparent, and that falls short of fully containing the complexity and uncertainty of life. Risk itself becomes mediated through software and becomes a processual output of normative values which are themselves the result of computational processes usually hidden within the lines of computer code. For example, software renders the display of financial portfolio information in a very stylised, simplified form, often with colour codings and increasingly with rich graphics.[6] Not only do few market participants fully understand risk as a statistical category, but the familiar bell-shaped curve of Gaussian distributions displayed on mobile screens, encourages a kind of 'domesticated' approach to risk that makes it appear familiarised and benign. Indeed, it is this misunderstanding of risk that Taleb (2007) blames for the huge leveraged asset bubble in 2007–2009 and the resultant financial crisis.[7] Indeed, only recently AXA S.A., the French financial services giant, was forced to reveal that 'that it had made a "coding error" that affected returns in its various portfolios in ways that had yet to be determined', but which could have resulted in substantial losses, and that '[i]t was an "inadvertent mistake" entered into one of AXA Rosenberg's main "risk models" by a computer programmer in April 2007' (Sommer 2010). Three years of a computer programming bug on a portfolio which at its height was worth $62 billion, demonstrates the profound effects that computer code can have, indeed, the portfolio, at the time of writing, is worth $41 billion after many investors have begun to leave the fund due to worries about the bug's effects.

These conceptualisations and arguments are clearly an important part of the content of financialisation, but now I would like to turn to the notion of the computational 'stream' by extending Lyotard cultural understanding of the stream. This concept helps to map real existing 'territories' (such as sensory, cognitive, affective and aesthetic) in relation to computational processes. Here software is active in the creation and maintenance of a temporal dimension that supports particular kinds of subjectivity. Indeed, this links with Thrift's (n.d.) notion of our having a 'minimal conscious perception which is held in place by all manner of systems and environments and sites that extend awareness' (n.d.: 3). Here, we can think of the external management of the internal perception of time that is linked to a form of Heideggerian angst towards a future event – sickness, old age, and so forth – which provides a new affective fuel source for capitalism. This is an anxiety maintained through a destabilising sense of the rapid passage of time, manifested through, for example, continual and inexplicable changes in commodity prices, stock valuations, asset price expansions and contractions – themselves fed as data streams to the processual subject. These are connected through a series of mechanisms to the body, and here I am thinking of a machinic notion drawn from Deleuze and Guattari (2004), or to an emotional response to the representation of the future body given through a series of visual images, such as actuarial graphs and charts (again connected to the notion of mobile spaces of risk or financialisation through devices such as the Apple iPhone). This 'full-on or full palette capitalism' (Thrift n.d.: xx) functions through the exploitation of forethought, where the aim is to produce a certain expectation and preparedness into which a desire is linked to the intensification of action.

This life stream is therefore a performative subjectivity highly attenuated to interactivity and affective response to an environment that is highly mediatised and deeply inscribed by computational datascapes. This helps to explain the kinds of active investor subjects that Governments seek to encourage through financial regulation such as annual tax renewal requirements, for example in Investment Saving Accounts (ISAs), a form of tax-free saving account in the UK, which require the accounts to be moved or reinvested every April; or in the Swedish Pension case outlined by Belfrage (2008) where the intention was to encourage over 50 per cent of pension savers to undertake continual asset management activities in relation to Swedish worker's pension portfolios invested in the Stock Market (the actual number of active traders turned out to be only 8 per cent in 2005) (Belfrage 2008: 289).[8] Clearly, the financialisation of society remains a work in progress.

So financialised code is a complex set of materialities that we need to think carefully about in turn. From the material experience of the financialised user of code, both trader and consumer, to the reading and writing of code, and then finally to the execution and experience of code as it runs on financial trading systems, we need to bring to the fore how code is a condition of possibility for a computational stream whether of financial news and data, or of a datastream cognitive support for everyday life.

Lifestreams

I now want to look at the practice of creating lifestreams, particularly through the example of Twitter. Twitter is a web-based microblogging service that allows registered users to send short status update messages of up to 140 characters to others (Herring & Honeycut 2009: 1). From a few messages per day, known as Tweets, in 2006 the service took off in popularity in 2009 and has grown to handle over 90 million messages per day in 2010 (Twitter 2010). Twitter works by encouraging the uploading and sharing of photographs, geodata tags, updates on what you are doing and so forth, this is transformed into a real-time stream of data that is fed back onto the web and combined with the updates of other people whose user-stream you 'follow'. It is helpful to,

> think about Twitter as a rope of information — at the outset you assume you can hold on to the rope. That you can read all the posts, handle all the replies and use Twitter as a communications tool, similar to IM — then at some point, as the number of people you follow and follow you rises — your hands begin to burn. You realize you cant hold the rope you need to just let go and observe the rope (Wiener, quoted in Borthwick 2009).

Although originally considered a marginal activity, Twitter, and similar microblogging services, have risen dramatically in use throughout the last few years. Particularly as politicians and the media have caught on to the unique possibilities generated by this rapid communicational medium. Designated as solipsistic and dismissed at first by the pundits, the growth in Twitter's use has meant that it can no longer be ignored and indeed it has become a key part of any communication strategy for politics, corporations and the media more generally. Twitter has evolved rapidly from a simple messaging service, to a form of real-time rolling news reporting on political and other events, from formal political meetings to protest actions. Political examples from the UK have included Gordon Brown's foray into Twitter defending the NHS over the issue of NHS 'death panels' (Toppling and Muir 2009). More recently, we have witnessed usage by the political class in the UK across the whole of the country (The Independent 2009); Damian McBride's emails to LabourList blogger Derek Draper, which were widely 'retweeted' by Twitter users (BBC 2009); and increasing concerns over the freedom of speech implications posed by the libel action against the *Guardian* reporting a parliamentary question about Trafigura regarding its relationship to exporting materials, which was widely 'retweeted' following an injunction to stop reporting on the incident (Dunt and Stephenson 2009). There is an increasing need for a cartography of both the production and empirical content of a number of these collaborative, streamed institutions and their recording of political events, power and interests. Institutions as diverse as Downing Street, the White House, Scotland Yard, The UK Parliament, INTERPOL, NATO, the Labour Party, and the Conservative Party have all recently instituted mechanisms for using these real-time computational services to supplement the limitations of better established communications procedures.

The conditions underpinning this shift, however, are not solely communicational. What marks these real-time stream sites is their creation by the active contributions of an epistemic community surrounding the 'owner'. These communities are typically marked by very loose ties, often no more than a 'screen-name' or even anonymous contributions to the site through updates. They also have the capacity to create a form of social contagion effect whereby ideas, media and concepts can move across these networks extremely quickly. Over the past ten years, we have witnessed an explosion of media forms made possible by the peer-to-peer technologies of the Internet (Atton 2004, Benkler 2007, Gauntlett 2009, Terranova 2004) transforming political institutions and their relationship to citizens (Coleman 2005; Chadwick 2007). As such, real-time streams presents an excellent opportunity for tracing the impact of computational real-time devices in everyday life and the way in which they capture the informal representations of issues with which contemporary communities are becoming increasingly concerned. It is possible that Twitter and other real-time streams both decentre social structures and expand the numbers involved.

Filled with constant updating, real time 'tweets', Twitter users disseminate affect, opinion-formation, and information in a very Tardian way. Twitter, and similar real-time stream services, collect data from both elites and non-elites and can be used to reconstruct knowledge of social and political events in an online real-time context. Examples include the real-time Twitter feeds following national political debates, World Cup football matches, fashion and culture events, and the presentation of prestigious prizes and awards, such as Baftas or Grammys. The attention of political, technology and media communities have been captured by the emergence of the 'real-time web' using Twitter and other services such as Facebook, Quora, Diaspora and Meebo. But as more people participate and subscribe to the services, the difficulties in negotiating a large and complex information resource becomes acute. The network effects combined with the vast amount of information flowing through the network are difficult for the user to understand.

Twitter therefore acts to facilitate a form of social communication by rapidly distributing information and knowledge across different streams.

Indeed, Twitter is made up of streams of data that constitute a 'now web [that is] open, distributed, often appropriated, sometimes filtered, sometimes curated but often raw' (Borthwick 2009). But it is the technology that makes up Twitter that is a surprising: a simple light-weight protocol that enables the fast flow of short messages,

> The core of Twitter is a simple transport for the flow of data — the media associated with the post is not placed inline — so Twitter doesn't need to assert rights over it. Example — if I post a picture within Facebook, Facebook asserts ownership rights over that picture, they can reuse that picture as they see fit. If I leave Facebook they still have rights to use the image I posted. In contrast if I post a picture within Twitter the picture is hosted on which ever service I decided to use. What appears in Twitter is a simple link to that image. I as the creator of that image can decide whether I want those rights to be broad or narrow (Borthwick 2009).

Increasingly, we are also seeing the emergence of new types of 'geo' stream, such as location,[9] which give information about where the user is in terms of GPS co-ordinates, together with mixed media streams that include a variety of media forms such as photos, videos and music. Location based services, such as Facebook Places, FourSquare and Gowalla, enable a user to capture GPS information in real-time, updating this as a data stream recording places, activities and life events to the Internet. It is even argued that we are seeing the emergence of a new communication layer for the web based on micro-messages and sophisticated search. As Borthwick explains, '[i]f Facebook is the well organised, pre planned town, Twitter is more like new urban-ism — its organic and the paths are formed by the users' (Borthwick 2009). But this is not just a communications channel, it is also a distributed memory system, storing huge quantities of information on individuals, organisations and objects more generally. The things that are 'collected' and updated by users into these streams is remarkable, for example one user: (i) 'collect[s] sugar levels everyday (like 6 times per day). This helps me to "understand" my metabolism, my diet and my stress levels'; (ii) 'calorie expenditure and effort during my workouts'; (iii) 'blood glucose level every 5 minutes through a continuous glucose monitor stuck in my gut'; (iv) '[and] track my sexlife at bedposted.com (duration, intensity, positions)' (quoted on FlowingData 2010). This is what Kevin Kelly has revealingly called the quantified self (Kelly 2010). This raises serious privacy issues, but also the cultural and social implications of living life in such a public way, mediated through the code that is enabling and supporting these services.

These new real-time streams and their relationships to both individuals, organisations, culture and society, let alone the state and politics, are still an emergent sphere of research. Many questions remain unanswered, not the least of which is who owns these huge data reservoirs and how will this data be used in the future. Indeed, Twitter recently turned over every tweet in its archive to the Library of Congress and now all tweets are archived automatically,

> every public tweet, ever, since Twitter's inception in March 2006, will be archived digitally at the Library of Congress. That's a LOT of tweets, by the way: Twitter processes more than 50 million tweets every day, with the total numbering in the billions (Adams 2010).

These streams are fascinating on a number of different levels, for example questions remain over the way in which national identity might be mediated through these computational forms in terms of an imagined community composed of twitter streams that aggregates institutions, people and even places.[10] Real-time streams offer some exciting potential in terms of cultural streams and movements as aggregations of data streams, real-time State representation through state institutions in a constellation of streams, and even national aggregates. Whether new political subjectivities are enabled through these streams, one is sure that the data will be captured and analysed as the capacity of these life-stream systems mature. This will be increasingly revealing for real-time polls, opinion formation Tardian analysis of social aggregates, and, of course governments and multinational corporations eager to monitor and manipulate the creators of these streams.

So far, I have mapped a number of different strands which coalesce around notions of aesthetics, affectivity, risk and processual subjectivity. Most importantly, I think I have tried to outline the value of Lyotard notion of the stream as a concept for developing our understanding of the computational subjectivity. I have only been able to outline some of the key areas of enquiry which I think are relevant to this, and there is clearly much work to be done in understanding the relationship between forms of computational temporality, subject positions and technological mediation and materiality. Additionally the highly visualised form of data representation that is increasingly used to express data in a qualitative form, together with the computational relationship with self raised by reflexive use of life-streams also raise important questions.

In the final section, I want to shift focus and consider the wider implications of thinking-streams, computer code and software.

Subterranean streams

Perhaps the first principle that one might consider with respect to computational devices is that they appear to encourage a search for simple solutions and answers to problems, and therefore a backlash against complexity (especially when they become screenic – in common with other mediums such as television, film and print). The solutions become mediated through technological proposals which themselves rely on computational notions such as computability, distributed processing, intensively recursive dynamics and computationally correct narrative strategies. The language of nature, politics, culture, society and economics becomes infused with computability to the extent that data flows outside of human consciousness and that in order to understand and act upon them, additional computational strategies are required (this is indeed the paradigm suggested through digital humanities and cultural analytics frameworks). This points towards an intensity of fast moving technological culture that privileges data streams over meaning, that is, an explosion of knowing-that rather than knowing-how – and here we might note the current political fascination with Twitter and similar social networking sites.

This could lead to a situation in which the user is unable to perceive the distinction between 'knowing-how' and 'knowing-that' relying on the mediation of complexity and rapidity of real-time streams through technology. This Heidegger would presumably describe this as dasein no longer being able to make its own being an issue for itself. Indeed, this may even point to a homogeneity of being in the digital 'age', as we become a being whose existence is mediated by identical computational processes. This would have grave implications for a distributed fragmentary subject relying on computational devices that are radically uncertain and opaque. Indeed, if these computational devices are the adhesives which fix the postmodern self into a patterned flow of consciousness (or even merely visualised data), an ontological insecurity might be the default state of the subject when confronted with a society (or association) in which unreadiness-to-hand is the norm for our being-in-the-world.

To return to the question from Sellars and reframe it: it still remains difficult to reconcile the homogeneity of the manifest image with the non-homogeneity of the scientific one, but we have to additionally address the unreadiness-to-hand of the computational image which offers the possibility of *partial reconciliation* through uncertain affordances. Additionally, the computational image in mediating a world of information, computation and process might inevitably transform the manifest image of meaning and complexity by disconnecting the possibility of familiarity from the referential totality and the subsequent reclassification of the personhood of dasein.[11] As we are inserting the computational image into the structure of everyday things, and therefore into the structure of our everyday life and its knowing-how, the deeper implications remain unclear and raise the need for a deeper understanding of the centrifugal force of the computational image. If the manifest world is the world in which dasein, 'came to be aware of [itself] as [being]-in-the world', in other words, where dasein encountered him/herself as dasein (Sellars 1962: 38), then the eclipse or colonisation by equipment that remains unready-to-hand and that fragments and destabilises the possibility of a referential totality would suggest that the manifest image, in so far as it pertains to man or woman, is now potentially a 'false' image and this falsity threatens dasein as it is, in an important sense, the being which no longer has this image of itself.[12] For poststructuralist writers such as Foucault, talking about certain structural conditions of possibility,

> If... [they]... were to disappear as they appeared, if some event of which we can at the moment do no more than sense the possibility... were to cause them to crumble, as the ground of Classical thought did, at the end of the eighteenth century, then one can certainly wager that man would be erased, like a face drawn in sand at the edge of the sea (Foucault 2002: 422).

This would represent the final act in a historical process of reclassification of entities from persons to objects – potentially, dasein becoming an entity amongst entities, an stream amongst streams – with challenging political and cultural implications for our ability to trace the boundary between the human and non-human.[13] This, of course, returns us to the questions raised at the beginning of the book regarding humanity's ontological precariousness. In allowing the computational to absorb our cognitive abilities, off-loading the required critical faculties that we presently consider crucial for the definition of a life examined, we pay a heavy price, both in terms of the inability of computational methods to offer any way of engaging with questions of being, but also in the unreadiness-to-hand that computational devices offer as a fragmentary mediation of the world. This is where the importance of digital *Bildung* becomes crucial, as a means of ensuring the continued capability of dasein to use intellect to examine, theorise, criticise and imagine. It may also raise the possibility of a new form of resistance for a dasein that is always at the limit of emancipation as the being that is constantly dealing with equipment that is radically unready-to-hand. It is, to attempt to consider the way in which computation enables what Turing called the 'super-critical mind', one that is apt at generating more ideas than it received, rather than the sub-critical mind (Latour 2004: 248):

> The majority of [human minds] appear to be "sub-critical"... An idea presented to such a mind will on average give rise to less than one idea in reply. A smallish proportion are super-critical. An idea presented to such a mind may give rise to a whole "theory" consisting of secondary, tertiary, and more remote ideas (Turing 1950: 454).

The future envisaged by the corporations, like Google, that want to tell you what you *should* be doing next (Jenkins 2010), presents knowledge as 'knowing that', which they call 'augmented humanity', I consider this as a model of humanity that is a-critical. Instead, we should be paying attention to how computation can act as a *gathering* to promote generative modes of thinking, both individually and collectively, through super-critical modes of thinking created through practices taught and developed through this notion of digital *Bildung*. This would, as Latour explains, 'require all entities, including computers, cease to be objects defined simply by their inputs and outputs and become again things, mediating, assembling, gathering' (Latour 2004: 248).

In this book, I have attempted to outline a groundwork for understanding, in the broadest possible sense, how 'one know one's way around' in a world that is increasingly reliant on computational equipment, but more maps are needed. Computational ontologies tend towards an understanding of the world which, whilst incredibly powerful and potentially emancipatory, cannot but limit the possibilities of thought to those laid within the code and software which runs on the tracks of silicon that thread their way around technical devices (sub-criticality). Understanding software is a key cultural requirement in a world that is pervaded by technology, and as Vico argued, as something made by humans, software is something that can and should be understood by humans. Indeed, this remains a project that is still to be fully mapped and has important consequences for the fragmentary way-of-being which continues to be desired throughout the socio-technical technicity that makes up the computational image.

In the spirit of Lyotard's expression of an aesthetics of disruption, however, I want to end the book with an elusive ought. This is an ought that is informed by a reading of *Aesop's Tales* through Michel Serres and his notion of the parasite (Serres 2007). The parasite is used not as a moral category, but in connection with an actor's strategic activities to understand and manipulate the properties of a network. Here, the parasite acts as interference, as processes that combine and mix together domains, for Serres it is this recombinant property of circulation networks rather than their general underlying patterns that is crucial to understand them. He explains:

> A human group is organized with one-way relations, where one eats the other and where the second cannot benefit at all from the first... The flow goes one way, never the other. I call this semiconduction, this value, this single arrow, this relation without a reversal of direction, 'parasitic' (Serres 2007: 5).
>
> The introduction of a parasite into the system immediately provokes a difference, a disquilibrium. Immediately, the system changes; time has begun (Serres 2007: 182).

For example, parasitic economic activities manipulate goods already available and subvert them from their original function. They are embedded in such a way as to make their removal either impossible or too expensive – reminiscent of the phrase 'too big to fail'. Finance capital and the equipment it deploys to assemble the markets that sustain it therefore acts to counteract the way in which investors look for liquidity and their ability to invest where they cannot get 'stuck', and from which they can withdraw at the smallest sign of trouble. Through parasitic technologies, users are constantly enticed back into the market, where they themselves intend to eat at the benefit of another. Here, the notion of the stream is intensified through the action of time within computational networks, literally the 'ticks' of network time which reflect the actions of millions of streams within the network and which cascade through the data streams that are threaded through the networks and chains of causality. As Serres argues:

> To parasite means to eat next to. Let us begin with this literal meaning. The country rat is invited by his colleague from town, who offers him supper. One would think that what is essential is their relation of resemblance or difference. But that is not enough; it never was. The relation of the guest is no longer simple. Giving or receiving, on the

rug or on the tablecloth, goes through a black box. I don't know what happens there, but it functions as an automatic corrector. There is no exchange, nor will there be one. Abuse appears before use. Gifted in some fashion, the one eating next to, soon eating at the expense of, always eating the same thing, the host, and this eternal host gives over and over, constantly, till he breaks, even till death, drugged, enchanted, fascinated. The host is not prey, for he offers and continues to give (Serres 2007: 7).

Aesop's fable ends with the country mouse returning home declaring that it is better to be able to enjoy what you have in peace, than live in fear with more. But, Serres (2007) also gestures towards an alternative parasitic understanding in his retelling of the fable. The question of who this subject 'eats next to', is perhaps reflected in the way in which streams pass through other streams, consumed and consuming, but also in the recorded moments and experiences of subjects who remediate their everyday lives. This computational circulation, mediated through real-time streams, offers speculative possibilities for exploring what we might call parasitic subjectivity. Within corporations, huge memory banks are now stockpiling these lives in digital bits, and computationally aggregating, transforming and circulating streams of data – literally generating the standing reserve of the postdigital age. Lyotard's (1999: 5) comment to the streams that flow through our postmodern cultural economies seems as untimely as ever: 'true streams are subterranean, they stream slowly beneath the ground, they make headwaters and springs. You can't know where they'll surface. And their speed is unknown. I would like to be an underground cavity full of black, cold, and still water'.

Notes

1 The Idea of Code

1. Another way of saying this would be: the delegation of inscription systems from paper and other physical materials to computer software. The chief difference being that the inscriptions are mobile, quick, mutable (rather than immutable) and can reflexively change their own content (data).
2. Hutchins (1996), for example, describes the evolution of a complex naval navigation system for steering a warship that demonstrates how cognition can be decentred and then slowly delegated to computational devices.
3. Relaxed stability aircraft are designed to deviate from controlled flight without constant input, this means that the aircraft is always on the verge of going out of control. In contrast most aircraft are designed with positive stability, which means that following a disturbance, such as turbulence, the aircraft will return to its original attitude. Flying a negative stability design is therefore much harder, and without the assistance of the fly-by-wire computer systems would be extremely taxing for a fighter pilot.
4. The use of software in life-critical systems raises serious concerns, discussed in detail in *Killed by Code: Software Transparency in Implantable Medical Devices* (Sandler 2010).
5. Sometimes referred to as 'de-materialisation', that is the transfer of a logic from a mechanical process or container to a representation within binary data on a computer system. This is usually stored as 0s and 1s on a magnetic storage device such as a computer hard drive but can also by optically stored as binary pits on an optical storage device such as a CD or DVD. This is the form of the embedded code that runs a great number of appliances.
6. An interesting collection of videos of papers demonstrating early work in this field is to be found at the Softwhere: Software Studies 2008 website, http://workshop.softwarestudies.com/
7. http://webscience.org/home.html
8. More examples can be seen in the collection at http://googlelolz.com/
9. Of course, having written the book in the UK, we should note that Google now automatically attempts to localise data for the search results too. Google 'knows' where I am and what users in a particular region, country or locale may be searching for at any particular moment. Google also supplies an online tool called Google Analytics that allows the user to monitor searches and perform analytics upon them.
10. The top search engines by volume of search in July 2010 were Google 71.31%, Yahoo 14.47%, Bing 10.03%, Ask 2.27%, Aolsearch 1.19%, Others 0.73% (Hitwise 2010).
11. Also interestingly when one types 'Google is' one is presented with 'Google is Skynet'. Skynet was the defence computer system that becomes conscious and eventually takes over the world and tries to kill all humans in the film *The Terminator*.

12. An example is given by Kelly (2006: 39): 'Already the following views are widespread: thinking is a type of computation, DNA is software, evolution is an algorithmic process. If we keep going we will quietly arrive at the notion that all materials and all processes are actually forms of computation. Our final destination is a view that the atoms of the universe are fundamentally intangible bits. As the legendary physicist John Wheeler sums up the idea: "Its are bits"'.
13. Thus for some theorists of computation, it is argued that the underlying fabric of the universe is radically discrete made up of individual cells upon which a certain number of operations can be carried out, but these cells frame the exteriority of the universe within which our universe 'runs'. In other words, as Fredkin argues, our universe consists of information processing running on this universal computer, and in effect the universal computer is running a program code similar to the cellular automata that computer scientists have been experimenting with since the 1970s, such as Conway's Game of Life. Researchers like Wolfram further assert that they have discovered a 'new science' and argue that cellular automata underlie and explain the complexity of living systems themselves, such as DNA replication, cells, or complete biological systems. In the Regime of Computation, 'code is understood as a discourse system that mirrors what happens in nature and that generate nature itself' (Hayles 2005: 27). This is an idealist notion of computation that is divorced from the materiality of the medium and hence: 'first of all, it doesn't matter what the information is made of, or what kind of computer produces it. The computer could be of the conventional electronic sort, or it could be a hydraulic machine made of gargantuan sewage pipes and manhole covers, or it could be something we can't even imagine. What's the difference? Who cares what the information consists of? So long as the cellular automaton's rule is the same in each case, the patterns of information will be the same, and so will we, because the structure of our world depends on pattern, not on the pattern's substrate; a carbon atom... is a certain configuration of bits, not a certain kind of bits' (Wright 1988).
14. "Analog computation... is a form of experimental computation with physical systems called analog devices or analog computers. Historically, data are represented by measurable physical quantities, including lengths, shaft rotation, voltage, current, resistance, etc., and the analog devices that process these representations are made from mechanical or electro-mechanical or electronic components... Here experimental procedures applied to the machine, especially measurements, play a special role. The inexactness of the measurement means that only an approximate input can be measured and presented to the analog device, and only an approximate output can be measured and returned from it" (Tucker and Zucker 2007: 2).
15. 'Oracles' come from the work of Alan Turing (1939) in *Systems of Logic defined by Ordinals*, where the question raised by Turing was regarding the impact on a formal system of supplementing uncomputable deductive steps. Turing 'defined the "oracle" purely mathematically as an uncomputable function, and said, "we shall not go further into the nature of this oracle apart from saying that is cannot be a machine." The essential point of the oracle is that it performs non-mechanical steps.' (Hodges 2000).
16. It should be noted that digital philosophy's explanation of the computational universe would include instrumental rationality as well as other forms

of rationality – such as communicative (defined by community and debate) and aesthetic (defined by sensitivity to affect). This is, of course, an implication of an ontology that claims that the universe is software running on a universal computer and therefore must encompass all aspects of rationality and human action. This raises questions regarding a mechanistic notion of the structure of the universe, but also point towards issues over determinism and prediction which I can only highlight here.
17. Clearly, the question of time in the computational image is a fundamental one which is outside the scope of this chapter, but time has to manifest itself both outside and inside of the fabric of the computational universe in some way. We might then think of computational time as succession governed by a universal clock in a synchronous process, although there is presumably no reason why it might not be asynchronous, and distributed. One might further argue that the computational image encourages a relationship with either the manifest or the scientific image that is based on discrete time, each step is in a sense computationally independent and atomistic.
18. This is a double 'disappearance' or appresentation of the object; in its original mediation and then internally within the data structures of the machine.
19. HTML is the HyperText Markup Language used to encode webpages. AJAX is shorthand for Asynchronous JavaScript and XML, which is a collection of client side technologies that enable an interactive and audio-visual dynamic web.
20. I am indebted to Alan Finlayson for his comments on this section.
21. For example in *The Idea of a University* (Newman 1996) and *Culture and Anarchy* (Arnold 2009).
22. See http://www.bcs.org/server.php?show=nav.5829
23. What Heidegger calls 'the Danger' (*die Gefahr*) is the idea that a particular ontotheology should become permanent, particularly the ontotheology associated with technology and enframing (see Heidegger 1993a).
24. This does not preclude other more revolutionary human-computer interfaces that are under development, including haptic interfaces, eye control interfaces, or even brain-wave controlled software interfaces.
25. See http://www.thecomputationalturn.com/
26. See the open digital humanities translation of Plato's *Protagoras* for a good example of a wiki-based project, http://openprotagoras.wikidot.com/

2 What Is Code?

1. Sharon Hopkins, a poet who writes in the computer language called PERL (practical extraction and report language), explains 'that perl poetry is the first effort to "develop human-readable creative writings in an existing programming language... that not only [have] meaning in [themselves] but can also be successfully executed by a computer."' (Black 2002: 142).
2. 'An indulgence is a remission before God of the temporal punishment due to sins whose guilt has already been forgiven, which the faithful Christian who is duly disposed gains under certain defined conditions through the Church's help when, as a minister of redemption, she dispenses and applies

with authority the treasury of the satisfactions won by Christ and the saints' (*Indulgentiarum Doctrina* 1) (quoted on Catholic 2004).
3. This moral dimension to coding, especially when linked to the notion of free libre and open source software is discussed in Berry (2008).
4. This wholeness of the programmer as a humanistic literary subject is threatened by new techniques in programming that attempt to shift into an industrial rather than a craft-based approach to programming. Witness how modularity, object-oriented programming, agile programming and other Taylorist methods have been used to turn the software process into something more like a pin factory, for example.
5. In many ways this is pointing to the fact that there is a dual development of 'code' as an abstract set of concepts and practices around programming, and particular 'codes' which are instantiations of this, either in different languages or different implementations in the same language.
6. For any computer system to function requires that existing social practices are captured, rationalised, restructured and formatted to enable the implementation and operation of a computer system (rather than the other way around).
7. Many new technologies have been created to help with this process, including libraries of reusable code (such as free, libre and open source software), application programming interfaces (APIs), and visual source editors and interface builders.
8. It should be noted that code generated within the work hours of the employer, are the property of the employer under current intellectual property laws.
9. This points to what Marx called *moralischer Verschleiss* ('moral depreciation' in the official translations) and what is more accurately referred to as the wearing out of the processes embedded in the machine (Marx 1990: 528). I would like to thank Tom Cheesman for his help in uncovering the original meaning and the translation of the German *moralischer Verschleiss*.
10. The question raised by software 'wearing out' or 'ageing' is very interesting (see Parnas 1994). This is very different to the idea of physical wear and tear and is in software is in fact much closer to the concept of *moralischer Verschleiss* that Marx introduces.
11. For a discussion of the difficulties of software preservation see Mathew *et al.* (2010).
12. http://www.computerhistory.org/ and http://www.nationalmediamuseum.org.uk, for a useful software history bibliography, see http://www.cbi.umn.edu/research/shbib.pdf
13. For a good overview of the FUD approach see http://en.wikipedia.org/wiki/Fear,_uncertainty_and_doubt
14. Project Mercury had a FORTRAN syntax error in its computer code such as DO I=1.10 (not 1,10). The comma/period mistake was detected in software used in earlier suborbital missions and fixed. It would have had more serious repercussions in subsequent orbital and moon flights if it had not been addressed.
15. There is much work to be done here on developing ways of talking about code and software without unreflexively taking up the technical language of computer science and therefore having only a descriptive vocabulary to discuss these changes. We need concepts on a the number of different levels

relevant to understanding code and this type of concept formation can help us think critically through and reflect upon code more carefully.
16. The term 'proprietary' analytically in this chapter. Both of the following forms are included: (1) computer software produced by private actors (e.g. individuals or corporations); and (2) code produced within public institutions (e.g., government departments). This is because in neither case is the source code released to the public or supported by open development processes.
17. In the case of CD technology, which uses Pulse Code Modulation (PCM), the chunks (bytes) are 16 bits wide, that is, they are able to represent only 65535 different values within the wave, that are sampled at 44,100 times per second. In translating between the external world and the internal symbolic representation, information is lost as the 65535 values are a grid, the digital data structure, placed over a smooth waveform. When translated, or played back through a digital-analogue converter, those with keen ears (and expensive audiophile equipment) are able to hear the loss of fidelity and digital artefacts introduced by errors in translation between the two (i.e. back from digital data structure to analogue sound).
18. Cloud computing is the idea of allowing the user to store their documents away from their physical machines in the 'Cloud' which is essentially large data-centres located around the world. This allows the user constant access to their documents wherever they might be located, provided they have some form of Internet connection. Cloud computing is increasingly seeing a move of applications online too, sometimes called software-as-a-service (SaaS), which furthers the move, or 'dematerialisation', of the computing experience online.
19. Net Neutrality is the principle that any two actors on a network should be able to connect to each other at a certain level of access. As the Internet is a distributed network, corporations, who control particular portions of the Internet as private networks, are tempted to discriminate against other people's data on their part of the network allowing their own data to move more quickly (i.e. to slow down the others data). Of course, if everyone where to do this then the entire Internet would grind to a halt.

3 Reading and Writing Code

1. Clean room development is used when organisations want to make sure that there is no contamination of intellectual property. The programmers who work on a project are given no access to previous versions of source code in order to re-engineer a project from scratch, often in response to copyright infringement action, or the potential for one.
2. There are a number of these computer based jokes, one of my favourites being an attack on the Pascal programming language written in 1982, by Ed Post, called 'Real Programmers Don't Use Pascal', http://www.ee.ryerson.ca/~elf/hack/realmen.html
3. These code snippets are drawn from http://www.dreamincode.net/forums/topic/38102-obfuscated-code-a-simple-introduction/
4. http://www.ioccc.org/2004/kopczynski.hint
5. http://www.ioccc.org/1984/laman.hint
6. http://www.ioccc.org/2004/arachnid.hint

7. http://www.ioccc.org/2001/cheong.hint
8. http://www.ioccc.org/2001/rosten.hint

4 Running Code

1. This Linux version of the source code for 'Hello, world!' is available at http://asm.sourceforge.net/intro/hello.html
2. This was rendered using xxd.
3. Note, in this case for ease of representation an ELF binary has been used. Although most operating systems support the use of the ELF object file format, technically the Mac OS X and the Window operating systems use their own binary format. They can however run emulators that would allow this binary file to execute.
4. In ancient mythology Procrustes ('the stretcher') was a bandit from Attica killed by Theseus. In Eleusis, Procrustes had a bed which he invited passersby to lie down in. When they did so, he either stretched them or cut off body parts to make them fit into the bed. Procrustes therefore attempted to reduce people to one standard size (Plutarch 2010).
5. Latour (1987) was particularly referring to the stability offered by paper to structured data, 'the first to sit at the beginning and at the end of a long network that generates what I will call immutable and combinable mobiles' (Latour 1987: 227). Whether one can create immutable mobiles in software is an interesting theoretical question to the extent that the materiality of software/data is extremely ephemeral. This helps explain the importance of cryptography and verification mechanisms in e-democracy systems.
6. See Everett *et al.* (2008: 898) which gives the 'mean ballot completion times (in seconds) by voting method' as a way of assessing different forms of completing ballot papers.
7. After evaluating trails of eVoting conducted in May 2007 trials, the UK Electoral Commission recommended there should be "no more pilots of electronic voting without a system of individual voter registration" and "significant improvements in testing and implementation" (Post 2009).
8. http://code.google.com/
9. The political economy of software for election systems will be an interesting requirement, especially if a monopoly situation arises (as would be expected) in a national electoral system. Essentially leaving the nation state open to the desire of the manufacture to encourage software updates in a similar manner to the rest of the software industry (see Campbell-Kelly 2004).
10. The kinds of documentation that are useful for programming include: requirements specification, flowcharts and diagrams, formal language specifications (e.g. UML, Z) and test/use cases. There is also documentation within the source code called 'comments' written by the programmers to help others understand the code.

5 Towards a Phenomenology of Computation

1. In this chapter I do not directly deal with the question of computational sociality except with reference to the notion of referential totality. This issue

of the interpersonal mediated through the technical devices discussed in the chapter would raise important theoretical questions about a social relationship (i.e. dasein-with) that was mediated through unreadiness-to-hand.
2. Here I would like to gesture towards the dyad of fabrication and annihilation that Heidegger examines through the notion of technical beings, and withdrawal as part of Gestell's (Enframing) challenging of Dasein (Heidegger 1993a, 1993b).
3. Of course, one may not fully understand the complexities of the internal data structures or the complex calculations involved in making the system work. Nonetheless, there is a recognition that something is happening behind the screen connecting it to a wider network of computational devices and data sources. One therefore knows where to situate one's phone, or how to mount the SatNav – for otherwise it will not function. One could think of this as a form of computational education or computational disciplining of the user.
4. Having no technical knowledge is a positive disadvantage in trying to use these technologies that can be very unforgiving and cryptic if basic notions are not understood, for example satnav booting from the memory card.
5. 'A situated simulation requires a broadband (3G) smartphone with substantial graphics capabilities, GPS-positioning features, accelerometer and electronic compass. In a situated simulation there is approximate identity between the users visual perception of the real physical environment and the users visual perspective into a 3D graphics environment as it is represented on the screen. The relative congruity between the real and the virtual is obtained by letting the camera position and movement in the 3D environment be determined by the positioning and orientation hardware. As the user moves in real space the perspective inside the virtual space changes accordingly' (INVENTIO-project n.d.)
6. See, for example, 'SixthSense' a wearable gestural interface that augments the physical world with digital information and uses hand gestures to interact, http://www.pranavmistry.com/projects/sixthsense/
7. I am therefore concerned with the *formal indication* of the computational. That is '[We must] make a leap and proceed resolutely from there!... One lives in a non-essential having that takes its specific direction toward completion from the maturing of the development of this having.... The evidence for the appropriateness of the original definition of the object is not essential and primordial; rather, the appropriateness is absolutely questionable and the definition must precisely be understood in this questionableness and lack of evidence' (Heidegger 1985: 34–35, quoted in Dreyfus 2001a).
8. 'Research has shown that multitasking can have some strange effects on learning' (Poldrack 2010). Some researchers now argue that technologies that promote constantly distracted multitasking create 'switch costs' which change the brain systems that are involved so that even if one can learn while multitasking, the nature of that learning is altered and becomes less flexible (this notion of changing brain patterns is called 'neuroplasticity') (Doidge 2007).
9. Here I am making a distinction between the equipmental form of tools, machinery, and such like, and the specific example of computation devices which I argue are a specific case of equipment that does not withdraw.
10. Heidegger uses the term Dasein in Being and Time, Division I to mean the human way of being, literally 'being-there' or Dasein. In Division II he is

more interested in particular human beings and he talks about 'a Dasein'. Thus Heidegger is not studying Dasein but Dasein's way of being (see Dreyfus 2001a: 14).
11. Here Sellars use of the term 'image' does not refer to the visual as such, rather to the conceptual framework that organises experience.
12. Here I am gesturing towards the notion that computation is a Weltanschauung (Worldview) within particular disciplinary groups.
13. In other words vicariousness is being-for-another.
14. It is important to note that conspicuousness is not broken down equipment. Heidegger defines three forms of unreadyness-to-hand: Obtrusiveness (*Aufdringlichkeit*), Obstinacy (*Aufsässigkeit*), and Conspicuousness (*Auffälligkeit*), where the first two are non-functioning equipment and the latter is equipment that is not functioning at its best (see Heidegger 1978, particularly footnote 1).
15. Here I am referring to the ability of the user to do different things at the same time (multitask), rather than a property of the technical device or operating system.
16. This is not to say that Apple, in particular, has not tried through the use of layering in the operating system which take out of the hands of developers the problem of interface element design, however, these forms of simplifications of user interface by their very nature create complexities in use elsewhere.
17. Here it is interesting to note Horkheimer and Adorno's (2006) position that: 'The hope that the contradictory, disintegrating person could not survive for generations, that the psychological fracture within it must split the system itself, and that human beings might refuse to tolerate the mendacious substitution of a stereotype for the individual – that hope is in vain' (Horkheimer and Adorno 2006: 64).
18. This is not to say that analogue technologies do not break down, the difference is the *conspicuousness* of digital technologies in contrast to the *obstinacy* or *obtrusiveness* of analogue technologies.
19. eBook readers use new technologies such as e-ink which provides a very stable viewing experience, albeit not yet at the resolution of paper, which attempts to mimic the book form very closely. Unsurprisingly, however, the temptation to build into these devices menuing systems, messaging, webbrowsers, annotation features, search engines, plus libraries of thousands of books, still creates ample opportunity for maximum viewer distraction.
20. Here I use screen essentialism to point towards an understanding of the interface, whether explicitly screen-based or on the surface of the object, as the privileged site for research. Here the screen is understood as unproblematically representing the inner state of the device or even that knowledge of the screen alone is sufficient for research without any recourse to a deeper notion of the technical layers which underlie it.
21. However, as Daniel Hourigan pointed out, Babushka dolls do have a finite limit and therefore do not have the kind of infinite multiplicity my point aims at.

6 Real-Time Streams

1. Nietzsche was the first German professor of philology to use a typewriter; Kittler is the first German professor of literature to teach computer programming (Kittler 1999: XXXI).

2. One might say that the first principle of the real-time stream is, following Daniel Paul Schreber's explanation of the recording of his thoughts whilst suffering from mental illness described in his book, *Memoirs of My Nervous Illness*, '*scilicet* – written-down' (Conner, 2008). This Kittler describes as a 'writing down system' or *Aufschreibesystem*.
3. One can use the example here of recommendation systems that proclaim 'people like you also like X', in other words constituting the subject as a member of a specific set, whether a voting block, market segment or pattern.
4. The differing access to these narratives will be a key location of political contestation in the coming decades as people claim a right to access their own narrative datascapes, such as health records.
5. It is also interesting to see how life itself is conceptualised discursively as a process of financial flows and investment strategies particularly in relation to pension planning and investment guidance (e.g. Child Trust Funds) (see Finlayson 2008) and through propagation through the media (e.g. British television programmes such as 'Location, Location, Location' and 'Property Ladder').
6. An interesting example of this presentation of computational risk is the iPhone application ASBOrometer which computes the risk factor of a particular UK location through the use of government data, see http://www.asborometer.com/
7. This notion of risk is also different from Beck's (2002) notion of the Risk Society that posits the idea of risk as a transnational phenomena that transforms the ability of individual states to predict events and consequences, for example the Chernobyl nuclear disaster. Instead, risk here is understood as a statistical predictive category for managing the future within finance capital and markets.
8. The profitability of these markets is, to some extent, linked to the active trading strategies of investors or the velocity of trades, indeed throughout the dot com bubble which burst in 2001, profits from speculative day-traders using web-based trading software grew even as the costs of trading were reduced. Here we might note a connection, for example, between the intensity of trading activity and the profitability of investment corporations and their transformation into a retail service industry (e.g. see Charles Schwab).
9. Geodata contains location-based information, usually sourced from the GPS satellites (General Positioning System).
10. Some notable example include Big Ben in London http://twitter.com/big_ben_CLOCK, Tower Bridge http://twitter.com/towerbridge, the Earthquake twitter account, which logs worldwide earthquake events http://twitter.com/earthquake, lowflying rocks, which logs near earth object that passes within 0.2AU of Earth http://twitter.com/lowflyingrocks, and the 32m telescope at Cambridge, http://twitter.com/32m ; there are also interesting examples of animals tweeting such as http://twitter.com/common_squirrel and http://twitter.com/wmpcsidogsmithy
11. Where the referential totality is itself actualised by a series of unstable affordances, the referential chains of meaning are constantly in play. This flow of environment and infrastructure raises interesting questions regarding stabilising an association, for example.

12. Or as an alternative formulation, has only the computational image of itself as a basis of its self-knowledge.
13. This is, of course, not to suggest that the analytical rejection of this boundary cannot be productive for research, (see Actor Network Theory more generally, and Latour 2005, in particular) but rather to draw attention to the strictly political implications of, for example, reclassifying POWs as Enemy Combatants, or the removal of human rights from certain groups of people.

Bibliography

ACE (n.d.) The Alliance for Code Excellence, retrieved 1/7/2010 from http://codeoffsets.com/
Adams, R. (2010) All your Twitter belongs to the Library of Congress, *Guardian*, retrieved 03/08/2010 from http://www.guardian.co.uk/world/richard-adams-blog/2010/apr/14/twitter-library-of-congress
Alvarez, R. M. and Hall, T. E. (2008) *Electronic Elections: The Perils and Promises of Digital Democracy*. Oxford: Princeton University Press.
Arendt, H. (1971) 'Martin Heidegger at Eighty', *New York Review of Books*, retrieved 11/05/2010 from http://www.nybooks.com/articles/archives/1971/oct/21/martin-heidegger-at-eighty/
Arnold, M. (2009) *Culture and Anarchy*. Oxford: Oxford University Press.
Arthur, C. (2010) 'Digital Economy Bill Rushed Through Wash-up in Late Night Session', *Guardian*, retrieved 14/03/2010 from http://www.guardian.co.uk/technology/2010/apr/08/digital-economy-bill-passes-third-reading
Atton, C. (2004) *An Alternative Internet: Radical Media, Politics and Creativity*. Edinburgh: Edinburgh University Press.
Baltimoremd (n.d.) Windows 2000 Source Code, retrieved 1/6/2010 from http://www.baltimoremd.com/content/win2000source.html
Bassett, C. (2007) *The Arc and the Machine: Narrative and New Media*. Manchester: Manchester University Press.
BBC (2004) Microsoft source code leaked out, *BBC News*, retrieved 1/3/2010 from http://news.bbc.co.uk/1/hi/technology/3484545.stm
BBC (2009) No 10 apology over 'slur' e-mails. The BBC News Website. Retrieved 19/11/09 from http://news.bbc.co.uk/1/hi/7994408.stm
Beckett, C. (2008) *Supermedia: Saving Journalism So It Can Save the World*. London: Wiley–Blackwell.
Beer, D. and Gane, N. (2004) 'Back to the Future of Social Theory: an Interview with Nicholas Gane'. *Sociological Research Online*, retrieved 10/02/09 from http://www.socresonline.org.uk/9/4/beer.html
Beggs, E., Costa, J. F. and Tucker, J. V. (2009) 'Physical Experiments as Oracles', *Bulletin of the EATCS*, No. 97, pp. 137–51, February 2009.
Belfrage, C. (2008) Towards 'Universal Financialisation' in Sweden?!, *Contemporary Politics* (special issue on 'The Global Politics of Finance Capitalism'), 14 (3): 277–96.
Beniger, J. R. (1989) *The Control Revolution: Technological and Economic Origins of the Information Society*. London: Harvard University Press.
Benkler, Y. (2002) 'Coase's penguin, or Linux and the nature of the firm', *The Yale Law Journal*, 112: 369–446.
Benkler, Y. (2003a) 'Freedom in the Commons, Towards a Political Economy of Information', 52, *Duke L.J.* 1245.
Benkler, Y. (2003b) 'The Political Economy of Commons', *Upgrade*, Vol. IV., No.3, June 2003.

Benkler, Y. (2004) 'Sharing Nicely: On Shareable Goods and the Emergence of Sharing as a Modality of Economic Production',. *The Yale Journal*, 114: 273, 274–358.
Benkler, Y. (2006) *The Wealth of Networks*. London: Yale University Press.
Berry, D. M. (2004) 'The Contestation of Code', *Critical Discourse Studies*, 1(1), 65–89.
Berry, D. M. (2008) *Copy, Rip, Burn: The Politics of Copyleft and Open Source*. London: Pluto Press.
Berry, D. M. and Moss, G. (2006) 'Free and Open-source Software: Opening and Democratising e-government's Black Box', *Information Polity*, 11 (2006) 21–34.
Beunza, D. and Muniesa, F. (2005) 'Listening to the Spread Plot', In *Making things Public: Atmospheres of Democracy*, Bruno Latour and Peter Weibel (eds) (2005). London: MIT Press, pp. 628–33.
Beunza and Stark (2004) 'Tools of the Trade: The Socio-technology of Aarbitrage in a Wall Street Trading Room', *Industrial and Corporate Change*, Vol. 13, No. 2: 369–400.
Biancuzzi, F. and Warden, S. (2009) *Mastermind of Programming: Conversations with the Creators of Major Programming Languages*, Sebastopol: O'Reilly.
BJS (2010) Data sorting world record: 1 terabyte, 1 minute, retrieved 27/07/2010 from http://scienceblog.com/36957/data-sorting-world-record-falls-computer-scientists-break-terabyte-sort-barrier-in-60-seconds/
Black, M. J, (2002) *The Art of Code*. PhD dissertation, University of Pennsylvania.
Black, R. (2003) E-voting: Democratic or dangerous?, retrieved 14/03/2010 from http://news.bbc.co.uk/1/hi/world/americas/3169706.stm
Black, R. (2010) CRU climate scientists 'did not withhold data', *BBC News*, retrieved 7/7/2010 from http://news.bbc.co.uk/1/hi/science_and_environment/10538198.stm
Blackbox Voting (2010) Black Box Voting - America's Election Watchdog Group, retrieved 14/03/2010 from http://www.blackboxvoting.org/
Blattner, W. (2006) *Heidegger's Being and Time*. London: Continuum.
Bloom, P. (n.d.) Computing Fantasies: Psychologically Approaching Identity and Ideology in the Computational Age, retrieved 14/3/10 from http://www.thecomputationalturn.com/
Bogoslaw, D. (2007) Big Traders Dive Into Dark Pools, Business Week, 3 October 2010, retrieved from http://www.businessweek.com/investor/content/oct2007/pi2007102_394204.htm
Boltanski, L. and Thévenot, L. (2006) *On Justification: Economies of Worth*. Oxford: Princeton University Press.
Bolter, J. D. and Grusin, R. A. (1999) *Remediation: Understanding New Media*. London: MIT Press.
Borgmann, A. (1999) *Holding on to Reality: The Nature of Information at the Turn of the Millenium*. Chicago: The University of Chicago Press.
Borgmann, A. (2010) Orientation in Technological Space, First Monday, Vol. 15, 6–7, retrieved from http://www.uic.edu/htbin/cgiwrap/bin/ojs/index.php/fm/article/view/3037/2568
Borthwick, J. (2009) Distribution ... now, THINK / Musings, retrieved 1/7/2010 from http://www.borthwick.com/weblog/2009/05/13/699/

Bratton, B. (2008) All Design is Interface Design, Softwhere: Software Studies 2008, Calit2, UC San Diego, video presentation, retrieved 18/10/2009 from http://emerge.softwarestudies.com/files/12_Benjamin_Bratton.mov

Bremner, C. (2009) Top French court rips heart out of Sarkozy internet law, *Times Online*, 11 June 2009, retrieved 13/03/2010 from http://technology.timesonline.co.uk/tol/news/tech_and_web/article6478542.ece

Broukhis, L. A., Cooper, S., Noll L. C., and Seebach, P. (2006) 19th International Obfuscated C Code Contest Rules, retrieved 10/07/2010 from http://www.ioccc.org/2006/rules.txt

Calandrino, J. A., Feldman, A. J., Halderman J. A., Wagner, D., Yu, H. and Zeller W. P. (2007) Source Code Review of the Diebold Voting System, retrieved 13/03/2010 from http://www.sos.ca.gov/elections/voting_systems/ttbr/diebold-source-public-jul29.pdf

Campbell-Kelly, M. (2004) *From Airline Reservations to Sonic the Hedgehog: A History of the Software Industry*. London: MIT Press.

Callon, M. (1998) *The Laws of Markets*. London: Blackwell.

Callon, M. (2007)' An Essay on the Growing Contribution of Economic Markets to the Proliferation of the Social', *Theory, Culture & Society*, vol. 24(7–8): 139–63.

Carr, N. (2008) 'Is Google Making Us Stupid?', *The Atlantic Magazine*, July/August 2008, retrieved 18/06/2010 from http://www.theatlantic.com/magazine/archive/2008/07/is-google-making-us-stupid/6868/

Carr, N. (2010a)' Googlethink: The Giant's Creepy Efforts To Read My Mind', *The Atlantic Magazine*, July/August 2010, retrieved 18/06/2010 from http://www.theatlantic.com/magazine/archive/2010/07/googlethink/8120/

Carr, N. (2010b) Steven Pinker and the Internet, retrieved 19/06/2010 from http://www.roughtype.com/archives/2010/06/steven_pinker_a.php

Carr, N. (2010c) 'The Web Shatters Focus, Rewires Brains', *Wired Magazine*, June 2010, retrieved 21/06/2010 from http://www.wired.com/magazine/2010/05/ff_nicholas_carr/all/1

Castells, C. (1996) *The Information Society: The Rise of the Network Society*. Oxford: Blackwell.

Catholic (2004) Primer on Indulgences, retrieved 18/10/2009 from http://www.catholic.com/library/Primer_on_Indulgences.asp

CEV (2006) Commission on Electronic Voting, Ireland, retrieved 13/03/2010 from http://www.cev.ie/index.htm

Chadwick, A. (2006) *Internet Politics*. Oxford: Oxford University Press.

Chadwick, A. (2007) 'Digital Network Repertoires and Organizational Hybridity', *Political Communication*, 24 (3): 283–301.

Chalmers, D. (1989) Analog vs. Digital Computation, retrieved 18/10/2009 from http://consc.net/notes/analog.html

Chopra, S. and Dexter, S. (2008) *Decoding Liberation: The Promise of Free and Open Source Software*. Oxford: Routledge.

Chun, W. H. K. (2008) 'On "Sourcery," or Code as Fetish', *Configurations*, 16:299–324.

Clement, T., Steger, S., Unsworth, J. and Uszkalo, K. (2008) How Not to Read a Million Books, retrieved 21/06/2010 from http://www3.isrl.illinois.edu/~unsworth/hownot2read.html#sdendnote4sym

Coleman, B. (2009) 'Code Is Speech: Legal Tinkering, Expertise, and Protest Among Free and Open Source Developers', *Cultural Anthropology*, August 2009, Vol. 24, issue 3, pp. 420–54.

Coleman, S. (2005) 'The Lonely Citizen: Indirect Representation in an Age of Networks', *Political Communication*, 22(2): 197–214.
Coleman, S. and Blumler, J. (2009) *The Internet and Democratic Citizenship: Theory, Practice and Policy*. Cambridge: Cambridge University Press.
Coleman, S., Donk, W. and Taylor, J. (1999) *Parliament in the Age of the Internet*. Oxford: Oxford University Press.
Conner, S. (2008) Scilicet: Kittler, Media and Madness, lecture given at Tate Modern, 28 June 2008, retrieved 7/7/2010 from http://www.bbk.ac.uk/english/skc/scilicet/
Cooper, J. M. (1997) *Plato: Complete Works*, Cambridge: Hackett.
Curtis, P. (2010) 'Voting System Rated Not Fit for Purpose', *Guardian*, retrieved 13/04/2010 from http://www.guardian.co.uk/uk/2010/mar/15/voting-system-not-fit-electoral-commission
Deleuze, Gilles (1992) 'What is a dispositif?', In Armstrong, T. J. (ed.), *Michel Foucault Philosopher*. New York: Routledge, pp. 159–68.
Deleuze, G. and Guattari, F. (2000) *Thousand Plateaus*. London: Continuum.
Digitalcraft.org (2006) Obfuscated C code, retrieved 13/04/2010 from http://www.digitalcraft.org/iloveyou/c_code.htm
Digital Economy Bill (2009) Digital Economy Bill [HL], retrieved 13/03/2010 from http://www.publications.parliament.uk/pa/ld200910/ldbills/001/10001.i-ii.html
Dix, A., Finlay, J., Abowd, G. D. and Beale, R. (2003) *Human Computer Interaction*. London: Prentice Hall.
Doel, M. (2009)' Miserly Thinking/Excessful Geography: From Restricted Economy to Global Financial Crisis. *Environment and Planning D: Society and Space*, doi:10.1068/d7307
Doidge, N. (2007) *The Brain That Changes Itself: Stories of Personal Triumph from the Frontiers of Brain Science*. New York: Viking.
Douglas, J. (2008) #Include Genre, Softwhere: Software Studies 2008, Calit2, UC San Diego, video presentation, retrieved 18/10/2009 from http://emerge.softwarestudies.com/files/11_Jeremy_Douglass.mov
Drahos, P. and Braithwaite, J. (2003). *Information Feudalism: Who Owns the Information Economy?* Norton.
Dreyfus, H. (1995) How Heidegger Defends the Possibility of a Correspondence Theory of Truth with respect to the Entities of Natural Science, retrieved 18/10/2009 from http://socrates.berkeley.edu/~hdreyfus/rtf/Heidegger-Realism_5_95.rtf
Dreyfus, H. (2001b) *Being-in-the-world: A Commentary on Heidegger's Being and Time, Division I*. USA: MIT Press.
Dreyfus, H. (2001b) *On the Internet*. London: Routledge.
Dubray, J. (2009) *On the Origins of Cognitive Science: The Mechanization of Mind*. London: MIT Press.
Dunt, I. and Stephenson, A. (2009) *Guardian* claims victory after Trafigura Twitter frenzy, *Guardian*, retrieved 19/11/09 from http://www.politics.co.uk/news/culture-media-and-sport/guardian-gagging-order-sparks-twitter-frenzy-$1333687.htm
Economist, The (2009) 'Rise of the Machines', retrieved 10/05/2010 from http://www.economist.com/node/14133802
Economist, The (2010a) 'System Error', retrieved 10/05/2010 from http://www.economist.com/world/asia/displaystory.cfm?story_id=16068922

Economist, The (2010b) 'Science Behind Closed Doors', retrieved 10/05/2010 from http://www.economist.com/node/16537628

Economist, The (2010c) 'Data, Data Everywhere', retrieved 11/05/2010 from http://www.economist.com/node/15557443

Economist, The (2010d) 'All Too Much', retrieved 11/05/2010 http://www.economist.com/node/15557421

Economist, The (2010e) 'Needle in a Haystack', retrieved 11/05/2010 http://www.economist.com/node/15557497

Economist, The (2010f) 'Handling the Cornucopia', retrieved 11/05/2010 http://www.economist.com/node/15557507

Ellul, J. (1973) *The Technological Society*. London: Random House.

Everett, S. P., Greene, K. K., Byrne, M. D., Wallach, D. S., Derr, K., Sandler, D. and Torous, T. (2008) Electronic Voting Machines versus Traditional Methods: Improved Preference, Similar Performance. CHI 2008, 5–10 April 2008, Florence, Italy, retrieved 13/03/2010 from http://chil.rice.edu/research/pdf/EverettGreeneBWDST_08.pdf

Finlayson, A. (2008) Characterising New Labour: The Case of the Child Trust Fund, *Public Administration*, Vol. 86, 1, 2008.

FlowingData (2010) Discuss: Why collect data about yourself?, retrieved 03/08/2010 from http://flowingdata.com/2010/07/30/discuss-why-collect-data-about-yourself/

Foucault, M. (2002) *The Order of Things*. London: Routledge Classics.

Freeman and Gelernter (1996) The Yale Lifestreams Project Page, Circa 1996, retrieved 8/4/2010 from http://cs-www.cs.yale.edu/homes/freeman/lifestreams.html

Frewin, J. (2010) Chaotic polling problems lead to calls for e-voting, retrieved 13/05/2010 http://news.bbc.co.uk/1/hi/technology/10102126.stm

Fuller, M. (2003) *Behind the Blip: Essays on the Culture of Software*. London: Autonomedia.

Fuller, M. (2006) Software Studies Workshop, retrieved 13/04/2010 from http://pzwart.wdka.hro.nl/mdr/Seminars2/softstudworkshop

Fuller, M. (2008) *Software Studies\A Lexicon*. London: MIT Press.

Fuller, S. (2006) *The New Sociological Imagination*. London: Sage.

Fuller, S. (2010) 'Humanity: The Always Ready – or Never to be – Object of the Social Sciences?', in Bonwel, J. W. (ed.), *The Social Sciences and Democracy*. London: Palgrave Macmillan.

Galloway, A. (2006) *Protocol: How Control Exists After Decentralization*. London: MIT Press.

Gane, N. (2003) 'Computerized Capitalism: The Media Theory of Jean-François Lyotard', *Information, Communication & Society*. 6:3: 430–50.

Garnham, N. (2005) 'From Cultural To Creative Industries: An Analysis of the Implications of the "Creative Industries" Approach to Arts and Media Policy Making in the United Kingdom', *International Journal of Cultural Policy*, Vol. 11, No. 1, pp. 15–29.

Gauntlett, D. (2009) 'Media Studies 2.0: a response', *Westminster Papers in Communication and Culture*, special issue on Media Studies 2.0, Vol. 1, No. 1, pp. 147–58.

Geere, D. (2010a) Tunable "Sound Cloud" alters acoustics at will, Wired, retrieved 1/7/2010 from http://www.wired.co.uk/news/archive/2010-06/14/tunable-sound-cloud-alters-acoustics-at-will

Geere, D. (2010b) Programmable origami folds itself into shape, Wired, retrieved 1/7/2010 from http://www.wired.co.uk/news/archive/2010-07/1/programmable-origami

Gibson, J. J. (1977) 'The Theory of Affordances', In R. E. Shaw and J. Bransford (eds), *Perceiving, Acting, and Knowing*. Hillsdale, NJ: Lawrence Erlbaum Associates.

Gill, R. and Pratt, A. (2008) 'The Social Factory? Immaterial Labour, Precariousness and Cultural Work', *Theory, Culture & Society*, December 2008, Vol. 25 Nos 7–8, 1–30.

Gillespie, T. (2008) The Politics of "Platforms", retrieved 11/7/2010 from http://web.mit.edu/comm-forum/mit6/papers/Gillespie.pdf

Gliestoel (2010) Sitsim Demo II, retrieved 11/7/2010 from http://www.youtube.com/watch?v=NliEGCnlSwM

Golumbia, D. (2009) *The Cultural Logic of Computation*. Harvard: Harvard University Press.

Google (2010a) What this is really about: keeping the Internet open for consumers, retrieved 13/04/2010 from http://googlepublicpolicy.blogspot.com/search/label/Net%20Neutrality

Google (2010b) 2010 Financial Tables: Income Statement Information, retrieved 13/04/2010 from http://investor.google.com/financial/tables.html

Greenwood, C. (2007) Air Force Looks at the Benefits of Using CPCs on F-16 Black Boxes, CorrDefence, retrieved 13/04/2010 from http://www.corrdefense.org/CorrDefense%20Magazine/Spring%202007/feature.htm

Guatarri, F. (1996) *Chaosophy: Soft Subversions*. New York: Semiotext(e).

Halavais, A. (2008) *Search Engine Society*. London: Polity.

Hansen, M. B. (2006) *New Philosophy for New Media*. London: MIT Press.

Hardt, M. And Negri, A. (2000) *Empire*. London: Harvard.

Harman, G. (2009) 'On Vicarious Causation', *Collapse*, No. II.

Harry (2009) READ ME for Harry's work on the CRU TS2.1/3.0 datasets, 2006–2009!, retrieved 10/06/2010 from http://www.anenglishmanscastle.com/HARRY_READ_ME.txt

Hayles, N. K. (2004) 'Print Is Flat, Code Is Deep: The Importance of Media-Specific Analysis', *Poetics Today*, 25(1): 67–90.

Hayles, N. K. (2005) *My Mother Was a Computer*. Chicago: Chicago University Press.

Hayles, N. K. (2007) 'Hyper and Deep Attention: The Generational Divide in Cognitive Modes', *Profession*, No. 13, pp. 187–99.

Heidegger, M. (1966). *Discourse on Thinking*. New York: Harper and Row.

Heidegger, M. (1978). *Being and Time*. London: Wiley–Blackwell.

Heidegger, M. (1988) *The Basic Problems of Phenomenology*. USA: Indiana University Press.

Heidegger, M. (1993a) 'The Question Concerning Technology', In Krell, D. F (ed.), *Martin Heidegger: Basic Writings*. London: Routledge, pp. 311–41.

Heidegger, M. (1993b) 'Letter on Humanism', In Krell, D. F (ed.), *Martin Heidegger: Basic Writings*. London: Routledge, pp. 217–64.

Heidegger, M. (2010) Logic: The Question of Truth. Gesamtausgabe, Band 21. Trans. Thomas Sheehand, Manuscript, retrieved 18/10/2009 from http://socrates.berkeley.edu/%7Ehdreyfus/185_f07/pdf/HeideggerHandout04Sept07.pdf

Heim, M. (1987) *Electric Language: A Philosophical Discussion of Word Processing*. London: Yale University Press.

Heim, M. (1993) *The Metaphysics of Virtual Reality*. Oxford: Oxford University Press.

Helmond, A. (2008) Video, slides and notes from my presentation on Software-Engine Relations at HASTAC II and SoftWhere 2008, retrieved 18/10/2009 from http://www.annehelmond.nl/2008/07/09/video-slides-and-notes-from-my-presentation-on-software-engine-relations-at-hastac-ii-and-softwhere-2008/

Hesmondhalgh, D. (2009) 'The Digitalisation of Music', In Pratt, A.C. and Jeffcut, P. (eds), *Creativity and Innovation in the Cultural Economy*. Abingdon and New York: Routledge.

Heusser, M. (2005) Beautiful Code, *Dr. Dobbs*, August 09, 2005, retrieved 1/7/2010 from http://www.drdobbs.com/184407802

Hitwise (2010) Top 20 Sites & Engines, retrieved 29/07/2010 from http://www.hitwise.com/us/datacenter/main/dashboard-10133.html

Hodges, A. (2000) Uncomputability in the work of Alan Turing and Roger Penrose, retrieved 18/06/2010 from http://www.turing.org.uk/philosophy/lecture1.html

Hofstadter, R. (1963) *Anti-Intellectualism in American Life*, USA: Vintage Books.

Hopkins, S. (n.d.) Camels and needles: computer poetry meets the perl programming language, retrieved 18/10/2009 from http://www.digitalcraft.org/iloveyou/images/Sh.Hopkins_Perl_Poetry.pdf

Honeycutt, C. and Herring, S. C. (2009). 'Beyond Microblogging: Conversation and Collaboration via Twitter', *Proceedings of the Forty-Second Hawai'i International Conference on System Sciences*. Los Alamitos, CA: IEEE Press.

Horkheimer, M. and Adorno, T. (2006) 'The Culture Industry: Enlightenment as Mass Deception', In Durham, Meenakshi G. and Kellner, D. (eds), *Media and Cultural Studies: Keyworks*. London: Blackwell.

HTCwire (2010) Algorithmic Terrorism on Wall Street, retrieved 06/08/2010 from http://www.hpcwire.com/blogs/Algorithmic-Terrorism-on-Wall-Street-100079719.html

Huber, W. (2008) Soft authorship, Softwhere: Software Studies 2008, Calit2, UC San Diego, video presentation, retrieved 18/10/2009 from http://emerge.softwarestudies.com/files/08_William_Huber.mov

Hughes, T. P. (2005) Human-built World: How To Think About Technology and Culture. Chicago: Chicago University Press.

Hutchins, E. (1996) *Cognition in the Wild*, US: MIT Press.

Hutchinson, A. (2009) Global Impositioning Systems: Is GPS technology actually harming our sense of direction?, *The Walrus*, November 2009, retrieved 7/7/2010 from http://www.walrusmagazine.com/articles/2009.11-health-global-impositioning-systems/2/

IBM (2008) First-of-a-Kind Technology to Help Doctors Care for Premature Babies, retrieved 18/10/2009 from http://www-03.ibm.com/press/us/en/press-release/24694.wss

Independent (2009) 'UK Politics on Twitter: A Regional Breakdown', *Independent*, retrieved 19/11/09 from http://www.independent.co.uk/news/uk/politics/uk-politics-on-twitter-a-regional-breakdown-1813466.html

Jenkins, H. W. (2010)' Google and the Search for the Future', *The Wall Street Journal*, 14 August 2010, retrieved from http://on.wsj.com/aippTz

Jones, T. (2004) Statement about the public distribution of windows source, retrieved 5/6/2010 from http://www.scribd.com/doc/2074843/Windows-Internals-Expert-Speaks-on-Source-Code-Leak-Updated
Kelly, K. (2010) The Quantified Self, retrieved 03/08/2010 from http://www.quantifiedself.com/
Kelty, C. K. (2008) *Two Bits: The Cultural Significance of Free Software and the Internet*. US: Duke University Press.
Kelly, K. (2006) 'The Computational Metaphor', In Hassan, R. and Thomas, J. (eds), *The New Media Theory Reader*, London: Open University Press.
Kennedy, H. (2010) 'Net Work: The Professionalisation of Web Design', *Media, Culture and Society*, Vol. 32, pp. 187–203.
KinsmanThoughts (2009) Debunking Climategate: The Source Code 2/2, *Youtube*, retrieved 10/4/2010 from http://www.youtube.com/watch?v=Bo3A4aIUg-Y
Kirschenbaum, M. (2004) 'Extreme Inscription: Towards a Grammatology of the Hard Drive', *TEXT Technology*, No. 2, pp. 91–125.
Kittler, F. (1987) 'Gramophone, Film, Typewriter', *October*, Vol. 41, Summer, pp. 101–18.
Kittler, F. (1997). *Literature, Media, Information Systems*, Johnston, J. (ed.). Amsterdam: OPA.
Kittler, F. (1999) *Gramophone, Film, Typewriter*. Stanford: Stanford University Press.
Knight, Frank H. (1971) [orig. 1921] *Risk, Uncertainty, and Profit*, with an introduction by George J. Stigler. Phoenix Books. Chicago: University of Chicago Press.
Knorr Cetina, K. and Bruegger, U. (2002) 'Global Microstructures: The Virtual Societies of Financial Markets',. *The American Journal of Sociology*, Vol. 107, No. 4, January, pp. 905–50.
Kohno, T., Stubblefield, A., Rubin, A. D. and Wallach, D. S. (2004) Analysis of an Electronic Voting System, retrieved 13/03/2010 from http://avirubin.com/vote.pdf
Kramer, S. (2006) 'The Cultural Techniques of Time Axis Manipulation: On Fredrich Kittler's Conception of Media', *Theory, Culture and Society*, Vol. 23(7–8), pp. 93–109.
Kuhn, T. S. (1996) *The Structure of Scientific Revolutions*. Chicago: Chicago University Press.
Kurniawan, S.H. and Zaphiris, P. (2001) 'Reading Online or on Paper: Which Is Faster?', Abridged Proceedings of the 9th International Conference on Human Computer Interaction, pp. 220–22 5–10 August, New Orleans, LA.
INVENTIO-project (n.d.) Situated Simulations: Designing a Mobile Augmented Reality Genre, retrieved 13/7/2010 from http://inventioproject.no/sitsim/
Lakatos, I. (1980) *Methodology of Scientific Research Programmes*, Cambridge: Cambridge University Press.
Langley, P. (2008) 'Sub-prime Mortgage Lending: A Cultural Economy', *Economy and Society*, 37:4, 469–94.
Latour (1986) 'Visualization and Cognition: Thinking with Eyes and Hands', *Knowledge and Society*, 6, pp. 1–40.
Latour, B. (1987) *Science in Action*. Cambridge, MA: Harvard University Press.
Latour, B. (1988) *The Pasteurization of France*. Cambridge, Massachusetts: Harvard University Press.

Latour, B. (1992) Where are the Missing Masses? Sociology of a Door, Retrieved 18/7/08 from http://www.bruno-latour.fr/articles/article/050.html
Latour, Bruno (2002) 'Gabriel Tarde and the End of the Social', In *The Social in Question: New Bearings in the History and the Social Sciences*, ed. Joyce, P.) London, Routledge, pp. 117–32.
Latour, B. (2004) 'Why has Critique Run out of Steam? From Matters of Fact to Matters of Concer', *Critical Inquiry*, 30, pp. 225–48.
Latour, B. (2005) *Reassembling the Social: An Introduction to Actor-Network-Theory*. Oxford: Oxford University Press.
Latour, B. (2010) 'Tarde's Idea of Quantification', In Candea, M. (ed.), *The Social After Gabriel Tarde: Debates and Assessments*. London: Routledge.
Lazer, D., A. Pentland, L. Adamic, S. Aral, A.-L. Barabási, D. Brewer, N. Christakis, N. Contractor, J. Fowler, M. Gutmann, T. Jebara, G. King, M. Macy, D. Roy and M. Van Alstyne (2009) 'Computational Social Science', *Science*, Vol. 323, Issue 5915, 6 February 2009, pp. 721–3.
Legon, J. (2004) Profanity, partner's name hidden in leaked Microsoft code, CNN, retrieved 1/1/2011 from http://articles.cnn.com/2004-02-13/tech/microsoft.source_1_mike-gullard-windows-code-source-code?_s=PM:TECH
Lessig, L. (1999) *Code and Other Laws of Cyberspace*, New York: Basic Books.
Lessig, L. (2002) *The Future of Ideas: the Fate of the Commons in a Connected World*. New York: Vintage.
Levy, P. (1999) *Collective Intelligence*, London: Perseus.
Levy, S. (2001) *Hackers: Heroes of the Computer Revolution*. London: Penguin.
LilB (2010) The Age Of Information MUSIC VIDEO DIRECTED BY LIL B, retrieved 1/7/2010 from http://www.youtube.com/watch?v=corY-FZAZog&feature=player_embedded
Lucas, R. (2010) 'Dreaming in Code', *New Left Review*, No. 62, March/April 2010, pp. 125–32.
Lyotard, J. F. (1984) *The Postmodern Condition: A Report on Knowledge*. Manchester: Manchester University Press.
Lyotard, J. F. (1993) 'A Postmodern Fable', *Yale Journal of Criticism*, 6:1, p. 237.
Lyotard, J. F. (1993) *The Inhuman: Reflections on Time*. London: Polity.
Lyotard, J. F. (1999) *Postmodern Fables*. USA: University of Minnesota Press.
Mackenzie, A. (2003) The problem of computer code: Leviathan or common power, retrieved 13/03/2010 from http://www.lancs.ac.uk/staff/mackenza/papers/code-leviathan.pdf
Mackenzie, A. (2006) *Cutting Code: Software and Sociality*, Oxford: Peter Lang.
Malik, O. (2009) Google May Buy Twitter. Or Not. But Why is Twitter So Hot?, Gigaom, retrieved 3/6/2010 from http://gigaom.com/2009/04/03/google-may-buy-twitter-or-not-but-why-is-twitter-so-hot/
Manovich, L. (2001) *The Language of New Media*. London: MIT Press.
Manovich, L. (2008) *Software takes Command,* retrieved 03/05/2010 from http://lab.softwarestudies.com/2008/11/softbook.html
Manovich, L. and Douglas, J. (2009) Visualizing Temporal Patterns In Visual Media: Computer Graphics as a Research Method, retrieved 10/10/09 from http://softwarestudies.com/cultural_analytics/visualizing_temporal_patterns.pdf
Marks, P. (2009) 'Net Piracy: The People vs the Entertainment Industry@, *New Scientist*, issue 2737, 3 December 2009, retrieved 13/03/2010 from

http://www.newscientist.com/article/mg20427375.200-net-piracy-the-people-vs-the-entertainment-industry.html

Marino, M. C. (2006) 'Critical Code Studies', http://www.electronicbookreview.com/thread/electropoetics/codology.

Martin, R. (2002) *Financialization of Daily Life*. US: Temple University Press.

Marx, K. (2004) *Capital*, London: Penguin.

Mathew, B., Shaon, A., Bicarregui, J. and Jones, C. (2010) ' Framework for Software Preservation', *The International Journal of Digital Curation*, Issue 1, Vol. 5, retrieved 20/6/2010 from http://www.ijdc.net/index.php/ijdc/article/viewFile/148/210

May, C. (2006)' Escaping the TRIPs' Trap: The Political Economy of Free and Open Source Software in Africa', *Political Studies*, Vol. 54, pp. 123–46.

Mayer, M. (2010) 'Regulating What Is "Best" in Search?, *Financial Times*, Thursday, July 15, 2010, retrieved 16/07/10 from http://www.ft.com/cms/s/0/0458b1a4-8f78-11df-8df0-00144feab49a.html

McChesney, R.W. (2007) *Communication Revolution: Critical Junctures and the Future of Media*. London: New Press.

McLuhan, M. (2001) *Understanding Media*. London: Routledge.

Merrin, W. (2009) 'Media Studies 2.0', In *Westminster Papers in Communication and Culture*, special issue on Media Studies 2.0, Vol. 1, No. 1, pp. 17–34.

Metcalfe, B. (2004) If Open-source Software is So Much Cooler, Why Isn't Transmeta Getting it?, retrieved 12/12/04, from http://www.infoworld.com/articles/op/xml/00/02/14/000214opmetcalfe.html

Minsky, H. (1992) The Financial Instability Hypothesis. Working Paper No. 74, retrieved 01/11/09 from http://levy.org/pubs/wp74.pdf

Misztal, B. (1996) *Trust in Modern Societies: The Search for the Bases of Social Order*, London: Polity.

Miwa, M. (n.d.a) The MATARISAMA, retrieved 01/11/09 from http://www.iamas.ac.jp/~mmiwa/XORensemble.html.

Miwa, M. (n.d.b) The Jaiken-Operation, retrieved 01/11/09 from http://www.iamas.ac.jp/~mmiwa/jaikenop.html.

Miwa, M. (2003a) Bolelo by Muramatsu Gear Engine for Orchestra played by New Japan Philharmonic Orchestra at Suntory Hall', Tokyo, retrieved 01/11/09 from http://www.iamas.ac.jp/~mmiwa/mgear.mov.

Miwa, M. (2003b) A definition of Reverse-Simulation Music founded on the three aspects of music, retrieved 01/11/09 from http://www.iamas.ac.jp/~mmiwa/rsmDefinition.html.

Miwa, M. (2003c) Formant Brothers "Ordering a Pizza de Brothers!", retrieved 01/11/09 from http://www.youtube.com/watch?v=FFvFlpVjEjM&feature=related.

Miwa, M. (2004a) Matarisama, performed by The Method Machine "The Computing Bodies", Yokohama, at the 11th Festival in Kanagawa Contemporary Arts Series. [Video]

Miwa, M. (2004b) "Jiyai Kagura", composed by members of the workshop "Making the imaginary folk entertainment", Sendai Mediatheque, Sendai. [Video]

Miwa, M. (2005a) Jaiken-bugaku. Performed by Time Travellers Ensemble. Exploration of Time at Yamaguchi Center for Arts and Media (YCAM), Yamaguchi. [Video]

Miwa, M. (2005b) Music for 'Hibi', performed by the members of a workshop at 'Possible Futures, Japanese postwar art and technology'. Intercommunication Center (ICC). Tokyo, retrieved 01/11/09 from http://www.youtube.com/watch?v=AWGZMuUHXP4.

Miwa, M. (2005c) Shaguma-sama, composed and performed by members of the workshop 'Folk Entertainment in the Future' at Yamaguchi Center for Arts and Media (YCAM). Yamaguchi (2005). [Video]

Miwa, M. (2006a) Jaiken-beats. Performed in 2006 at the Computing Music IV conference in Cologne, 2004. [Video]

Miwa, M. (2006b) '369' homage for Mr. B., played by the New Japan Philharmonic Orchestra at Suntory Hall, Tokyo. [Video]

Miwa, M. (2007) Reverse-Simulation Music, Cyber Arts 2007, Prix Ars Electronica. [DVD]

Miwa, M. (2009) "Le Tombeau de Freddie / L'Internationale" by Formant Brothers, retrieved 18 June 2010 from http://www.youtube.com/watch?v=hkfrU-EOQ-E

Miwa, M. (2010a) "NEO DO-DO-I-TSU" Formant Brothers (Part 1: Introduction), retrieved 18 June 2010 from http://www.youtube.com/watch?v=qrKQ-7BjubE&feature=player_embedded

Miwa, M. (2010b) "NEO DO-DO-I-TSU" Formant Brothers (Part 2: Performance), retrieved 18 June 2010 from http://www.youtube.com/watch?v=Gvok DEEHujQ&NR=1

Montfort, N. (2008) My Generation About Talking, Softwhere: Software Studies 2008, Calit2, UC San Diego, video presentation, retrieved 18 /10/2009 from http://emerge.softwarestudies.com/files/14_Nick_Montfort.mov

Montfort, N. (2009) *Expressive Processing*. London: MIT Press.

Montfort, N. and Bogost, I. (2009) *Racing the Beam: The Atari Video Computer System*, London: MIT Press.

Moretti, F. (2007) *Graphs, Maps, Trees: Abstract Models for a Literary History*, London, Verso.

Morrison, K. (1997) *Marx, Durkheim, Weber: Formations in Modern Social Thought*. London: Sage.

Mosco, V. (ed.) (1988) *The Political Economy of Information*, London: University of Winsconin.

Mosco, V. (2009) *The Political Economy of Communication*. London: Sage.

Nature (2007) 'A Matter of Trust, 449, pp. 637–38, 11 October.

Newman, J. H. (1996) *The Idea of a University*. Yale: Yale University Press.

New York Times (2010) A Multitasker's Perspective, 6 June 2010, retrieved 18/06/2010 from http://www.nytimes.com/interactive/2010/06/06/business/kord-pano.html

New York Times (2010) The Google Algorithm, 14 July 2010, retrieved 16/07/2010 from http://www.nytimes.com/2010/07/15/opinion/15thu3.html?_r=2

Niccolai, J. (2010) 'I/o promises new kind of containerised data centre', *Techworld*, retrieved 24/07/2010 from http://news.techworld.com/data-centre/3232369/i-o-promises-new-kind-of-containerised-data-centre/?olo=rss

Noll, L. C., Cooper, S., Seebach, P. and Broukhis, L. A. (2009) The International Obfuscated C Code Contest: The IOCCC FAQ, retrieved 1/06/2010 from http://www.ioccc.org/faq.html

Norris, P. (2001) *Digital Divide: Civic Engagement, Information Poverty, and the Internet Worldwide*. Cambridge: Cambridge University Press.

Open Access (n.d.) What is Open Access?, retrieved 13/03/2010 from http://www.eprints.org/openaccess/

Ordinateurs-de-Vote (2010) Ordinateurs-de-Vote.org, Citoyens et informaticiens pour un vote vérifié par l'électeur, retrieved 14/04/2010 from http://www.ordinateurs-de-vote.org/

O'Reilly, T. (2005a) What Is Web 2.0: Design Patterns and Business Models for the Next Generation of Software, retrieved 18/06/2010 from http://oreilly.com/web2/archive/what-is-web-20.html

O'Reilly, T. (2005b) Web 2.0: Compact Definition?, October 1, 2005, retrieved 18/062010 from http://radar.oreilly.com/archives/2005/10/web-20-compact-definition.html

Oram, A. and Wilson, G. (2007) *Beautiful Code*. London: O'Reilly.

ORG (2007a) May 2007 Election Report, Findings of the Open Rights Group Election Observation Mission in Scotland and England, retrieved 14/04/2010 from http://www.openrightsgroup.org/wp-content/uploads/org_election_report.pdf

ORG (2007b) Electronic Voting. A challenge to democracy?, retrieved 14/04/2010 from http://www.openrightsgroup.org/wp-content/uploads/org-evoting-briefing-pack-final.pdf

ORG (2010) Open Rights Group, retrieved 14/04/2010 from http://www.openrightsgroup.org/

Outhwaite, W. (2009) 'How much capitalism can democracy stand (and vice versa)?', *Radical Politics Today*, May 2009, retrieved 14/03/2010 from http://doiop.com/outhwaite

Parikka, J. (2007) *Digital Contagions: A Media Archaeology of Computer Viruses*, London: Peter Lang.

Parnas, D. L. (1994) 'Software Aging', International Conference on Software Engineering, *Proceedings of the 16th international conference on Software engineering*: 279–87.

Pearce, F. (2010) 'ockey Stick Graph Took Pride of Place in IPCC Report, Despite Doubts', *Guardian*, Tuesday 9 February 2010, retrieved 5/6/2010 from http://www.guardian.co.uk/environment/2010/feb/09/hockey-stick-graph-ipcc-report

Perelman, M. (2002) *Steal This Idea: Intellectual Property Rights and the Corporate Confiscation of Creativity*. London: Palgrave.

Pingdom (2010) Google, undisputed heavyweight champion of mobile search, retrieved 29/07/2010 from http://royal.pingdom.com/2010/07/29/google-undisputed-heavyweight-champion-of-mobile-search/

Plutarch (2010) Life of the Theseus, retrieved 14/03/2010 from http://www.theoi.com/Text/PlutarchTheseus.html

Poldrack, R. A. (2010) 'Addictive Signals', *New York Times*, retrieved 19/06/2010 from http://roomfordebate.blogs.nytimes.com/2010/06/07/first-steps-to-digital-detox/

Post (2009) E-Democracy, Parliamentary Office for Science and Technology POSTnote, retrieved 14/03/2010 from http://www.parliament.uk/documents/post/postpn321.pdf

Prosser, A. and Krimmer, R. (2004) Electronic Voting in Europe – Technology, Law, Politics and Society, Workshop of the ESF TED Programme together with GI and OCG Proceedings, retrieved 13/03/2010 from http://citeseerx.ist.psu.edu/viewdoc/download?doi=10.1.1.92.9091&rep=rep1&type=pdf#page=82

Project Canvas (2010) Home, Project Canvas, retrieved 01/07/2010 from http://www.projectcanvas.info/

Pryke, M. (2006) Speculating on geographies finance, retrieved 14/3/08 from http://www.cresc.ac.uk/documents/papers/wp24.pdf

Raley, R. (2008) The Time of Codework, Softwhere: Software Studies 2008, Calit2, UC San Diego, video presentation, retrieved 18 Oct 2009 from http://emerge.softwarestudies.com/files/16_Rita_Raley.mov

Raymond, E. (2009) 'Hiding the Decline: Part 1 – The Adventure Begins', *Armed and Dangerous*, retrieved 3/7/2010 from http://esr.ibiblio.org/?p=1447

Readings, B. (1996) *The University in Ruins*, London: Harvard University Press.

Reisinger, D. (2009) Finland makes 1Mb broadband access a legal right, *CNet News*, 14 October retrieved 13/03/2010 from http://news.cnet.com/8301-17939_109-10374831-2.html

Richtell, M. (2010)' Hooked on Gadgets, and Paying a Mental Price', *New York Times*, 7 June 6, retrieved 18 June 2010 from http://www.nytimes.com/2010/06/07/technology/07brain.html

Ross, A. (2008) 'The New Geography of Work: Power to the Precarious?', *Theory, Culture & Society*, December, Vol. 25, Nos 7–8, 31–49.

Rumsey, E. (2009)' Did Salman Rushdie envision the Web in 1990?', *Seeing the Picture*, retrieved 4/5/2010 from http://blog.lib.uiowa.edu/hardinmd/2009/05/13/did-salman-rushdie-envision-the-web-in-1990/

Sandler, J. (2010) Killed by Code: Software Transparency in Implantable Medical Devices, retrieved 29/07/2010 from http://www.softwarefreedom.org/resources/2010/transparent-medical-devices.html

Sandler, D., Derr, K. and Wallach, D. S. VoteBox: a tamper-evident, verifiable electronic voting system, retrieved 13/03/2010 from http://www.usenix.org/events/sec08/tech/full_papers/sandler/sandler_html/index.html

Schloz, T. (2008) 'Market Ideology and the Myths of Web 2.0', *First Monday*, Vol. 13, No. 3 - 3 March, retrieved 13/03/2010 from http://firstmonday.org/htbin/cgiwrap/bin/ojs/index.php/fm/article/viewArticle/2138/1945

Schonfeld, E. (2009) 'Jump Into The Stream', *TechCrunch*, 17 May, retrieved 1/7/2010 from http://techcrunch.com/2009/05/17/jump-into-the-stream/

Schreibman, S., Siemans, R. and Unsworth, J. (2008) *A Companion to Digital Humanities*. London: Wiley–Blackwell.

Scratch (n.d) 'Scratch, Imagine, Program, Share', http://scratch.mit.edu/.

Sell, S.K. (2003) *Private Power, Public Law: The Globalization of Intellectual Property Rights*. Cambridge: Cambridge University Press.

Sellars, W. (1962) Philosophy and the Scientific Image of Man, In Colodny, Robert (ed.) *Frontiers of Science and Philosophy*, Pittsburgh: University of Pittsburgh Press, pp. 35–78.

Selznak (2004) We Are Morons: a quick look at the Win2k source, *Kino5hin*, Monday, 16 February2004, retrieved 7/7/2010 from http://www.kuro5hin.org/story/2004/2/15/71552/7795

Serres, M. (2007) *The Parasite*. London: University of Minnesota Press.

Shunmugam, V. (2010) Financial markets regulation: The tipping point, retrieved from http://www.voxeu.org/index.php?q=node/5056

Sieglar, M. G. (2010) Google Revenue Up 24% For The Year, But Only Slightly For The Quarter As Paid Clicks Fell, retrieved 16/07/2010 from http://techcrunch.com/2010/07/15/google-q2-2010/

Silver, D. (2008) 'History, Hype, and Hope: An Afterward', *First Monday*, Vol. 13, No. 3, 3 March, retrieved 16/03/2010 from http://firstmonday.org/htbin/cgiwrap/bin/ojs/index.php/fm/article/viewArticle/2143/1950

Simondon, Gilbert (1980) [orig. 1958] *On the Mode of Existence of Technical Objects*, Mellamphy, N. (trans.). Paris: Aubier, Editions Montaigne.

Slashdot (2004) 'Windows 2000 & Windows NT 4 Source Code Leaks', *Slashdot*, retrieved 5/4/2010 from http://slashdot.org/article.pl?sid=04/02/12/2114228

Smythe, D. W. (2006) 'On the Audience Commodity and Its Work', In Kellner, D. and Durham, M. G. (eds), *Media and Cultural Studies: Keyworks*. London: Blackwell.

Sommer, J. (2010) 'The Tremors From a Coding Error', *New York Times*, 18 June, retrieved 15/07/2010 from http://www.nytimes.com/2010/06/20/business/20stra.html?partner=rss&emc=rss

Spivak, N. (2009) Welcome to the Stream – Next Phase of the Web, retrieved 1/6/2010 from http://novaspivack.typepad.com/nova_spivacks_weblog/2009/05/is-the-stream-the-next-new-metaphor.html

Stallman, R. M. (2002) *Free Software, Free Society: Selected Essays of Richard M. Stallman*. Boston: GNU Press.

Sterling, B. (2010) Atemporality for the Creative Artist, *Wired*, retrieved 1/7/2010 from http://www.wired.com/beyond_the_beyond/2010/02/atemporality-for-the-creative-artist/

Stickney, D. (2008) 'Charticle Fever', *American Journalism Review*, retrieved 18/03/2010 from http://www.ajr.org/Article.asp?id=4608

Stiegler, B. (1998) *Technics and Time: The Fault of Epimetheis*. Stanford: Stanford University Press.

Stiegler, B. (2007) 'The Discrete Image', In Derrida, J. and Stiegler, B. (eds), *Echographies of Television*. London: Polity.

Sussman, G. (1997) *Communication, Technology, and Politics in the Information Age*, London: Sage.

Taleb, N. (2008) *The Black Swan: The Impact of the Highly Improbable*. London: Penguin.

Terranova, T. (2007) 'Futurepublic: On Information Warfare, Bio-racism and Hegemony as Noopolitics', *Theory, Culture & Society*, Vol. 24(3), pp. 125–45.

Terras, M. (2010) Present, Not Voting: Digital Humanities in the Panopticon, retrieved 10/7/2010 from http://melissaterras.blogspot.com/2010/07/dh2010-plenary-present-not-voting.html

Thaler, R. H., and Sunstein, C. R. (2009) *Nudge: Improving Decisions About Health, Wealth and Happiness*. London: Penguin.

The Invisible Committee (2009) *The Coming Insurrection*, London: Semiotext(e).

The Matrix (1999) Directed by Andy Wachowski. USA, Groucho II Film Partnership. [Film].

Thompson, J. B. (1995) *Media and Modernity: A Social Theory of the Media*. London: Polity.

Thomson, I. (2003) 'Heidegger and the Politics of the University', *Journal of the History of Philosophy*, Vol. 41, No. 4, pp. 515–42.

Thomson, I. (2009) 'Understanding Technology Ontotheologically, or: The Danger and the Promise of Heidegger, an American Perspective, In Jan-Kyrre Berg Olsen, Evan Selinger, and Søren Riis (eds), *New Waves in the Philosophy of Technology*. New York: Palgrave Macmillan, pp. 146–66.

Thrift, Nigel (n.d) Re-inventing Invention. *The Generalization of Outsourcing and Other New Forms of Efficacy under Globalization*, retrieved 18/7/08 from http://www.gold.ac.uk/media/thrift.pdf

Toppling, A. and Muir, H. (2009) 'Gordon Brown Joins Twitter Campaign Defending NHS', *Guardian*, retrieved 19/11/09 from http://www.guardian.co.uk/society/2009/aug/13/stephen-hawking-nhs-twitter-welovethenhs

Trechsel, A. and Mendez, F. (2005) *The European Union and e-voting: Addressing the European Parliament's Internet Voting Challenge*, London: Routledge.

Tucker, J.V. and Zucker, J.I. (2007) 'Computability of Analog Networks', *Theoretical Computer Science*, 371, 115–46

Turing, A. M. (1939) Systems of Logic defined by Ordinals, PhD thesis, retrieved 13/06/2010 from http://plms.oxfordjournals.org/cgi/reprint/s2-45/1/161.pdf

Turing, A. M. (1950) 'Computing Machinery and Intelligence', *Mind*, October, pp. 433–60.

Twitter (2010) Tweets per day, retrieved 1/08/2010 from http://www.flickr.com/photos/twitteroffice/4990581534/sizes/l/in/photostream/

Ullman, E. (2004) *The Bug*, London: Anchor Books.

Votebox (2009a) IAuditoriumParams.java, retrieved 13/03/2010 from http://code.google.com/p/votebox/source/browse/trunk/votebox/IAuditoriumParams.java

Votebox (2009b) ChallengeEvent.java, retrieved 13/03/2010 from http://code.google.com/p/votebox/source/browse/trunk/votebox/VoteBox.java

Votebox (2009c) VoteBox.java, retrieved 13/03/2010 from http://code.google.com/p/votebox/source/browse/trunk/votebox/ChallengeEvent.java

Votebox (2009d) The VoteBox Electronic Voting System, retrieved 13/03/2010 from http://votebox.cs.rice.edu/

XcottCraver (2008) The 2008 Underhanded C Contest, retrieved 01/05/2010 from http://underhanded.xcott.com/?p=8

Waldrip-Fruin (2009) *Expressive Processing: Digital Fictions, Computer Games, and Software Studies*, London: MIT Press.

Wark, M. (2007) *Gamer Theory*. Boston: Harvard University Press.

Weber, M. (2002) *The Protestant Ethic and the Spirit of Capitalism*. London: Routledge.

Weber, S. (2005) *The Success of Open Source*, Boston: Harvard University Press.

Weiner, L. R. (1994) *Digital Woes*. New York: Addison Wesley.

Weizenbaum, J. (1984) *Computer Power and Human Reason: From Judgement to Calculation*. London: Penguin Books.

Williams, R. (2003) *Television: Technology and Cultural Form*. London: Routledge.

Wilson, D. (2010) Google nabs patent to monitor your cursor movements, retrieved 29/07/2010 from http://www.techeye.net/internet/google-nabs-patent-to-monitor-your-cursor-movements#ixzz0v4FeFXNJ

Winner, L. (2001) *Autonomous Technology: Technics-out-of-control as a Theme in Political Thought*. London: MIT Press.

wijvertrouwenstemcomputersniet (2009) The Netherlands return to paper ballots and red pencils, retrieved 13/03/2010 from http://wijvertrouwenstemcomputersniet.nl/English

Wright, R. (1988) 'Did the Universe Just Happen?', *The Atlantic Monthly*, Vol. 261, No. 4, p. 29.

Zuboff, S. (1988) *In the Age of the Smart Machine: The Future of Work and Power*, New York: Basic Books.

Index

\# include, 40

Advertising, 7, 99
Affective, 158
Affordance, 15, 132, 136, 140, 168
Agencement, 157
Agile programming, 45
Aiken, Howard, 47
AJAX, 18
Algorithm, 7, 100, 159
Alpha, 41, 67
Analogue computing, 12
Analytical engine, 4
Application programming interface(s), 15, 7
Arnold, Matthew, 19
Assembly language, 45
Atanasoff, John, 47
Atemporal, 150
Audience commodity, 7
Augmented reality, 124
Autonomy, 9

Babbage, Charles, 47
BBC, 50, 71
Behavioural marketing, 8
Beta, 41, 67
Bildung, 19, 20, 26, 168, 169
Black box, 15, 41, 67, 137
Born digital, 25
Bug, 45, 69

C++, 35
Capitalism, 162
Carr, Nicholas, 120
Circumspection, 124
Click-stream, 7
Climategate, 73
Clock, 12, 97, 104
Cloud computing, 56, 99
Code, 2, 9, 17, 31, 33, 36, 43, 52, 61, 62, 64, 66, 75, 92, 104, 136, 149, 161, 167, 169
  Aesthetics, 49
  As container, 50
  As engine, 46
  As image or picture, 48
  As medium of communication, 49
  Assembly, 95
  Beautiful, 48
  Binary, 96
  Commentary, 51, 54, 74, 100
  Critical, 51, 53
  Delegated, 51, 52, 101
  Ethnography, 94
  Everyday, 97
  Illiterate, 82
  Literate, 29, 82
  Object, 51, 55, 103, 104
  Prescriptive, 51, 53, 103, 104, 105, 137
  Running, 94, 97, 107, 114, 117, 123
  Spatiality, 98
Codebase, 41
Codex, 54
Code libraries, 9
Code work, 32, 37, 40
Cognition, 141
Communication, 13, 49, 117
Compiler, 44
Computable, 16
Computation, 2, 7, 10, 125, 127
Computational, 21, 145, 171
  Humanities, 21
Computational image, 131, 141, 168, 169
Computational turn, 23
Computationalism, 11
Computationality, 10, 22, 27, 129, 169
Computational knowledge society, 3
Computational rationality, 13
Computer code, 5, 104
Computer science, 10, 14, 21, 44, 64, 114, 129
Copyright, 61
Correlational induction, 131

Craft, 82
Creative industries, 5
Critical code studies, 4, 113
Cryptography, 110
Culture, 19
Cultural analytics, 4, 24, 167
Cultural software, 6, 17, 32
Cultures of software, 17
Cyberculture, 18
Cybernetics, 136, 147
Cyberspace, 153

Dark pools, 160
Dasein, 130, 168, 169
Data, 1, 6, 51, 165
    Visualisation, 24, 26, 48, 154, 155
Database, 26, 50, 120, 128, 141, 142
Dataflow, 6
Dataspace, 153
Data centre, 50
Data-mine, 2
Debugger, 38
DeCSS, 53
Derrida, 11
Difference engine, 47
Digital, 20
    Bildung, 20
    Literacy, 20
    Stream, 51, 53, 55
Digital culture, 17
Digital data structure, 51, 54, 95, 107
Digital divide, 22, 113
Digital humanities, 18, 23, 167
Digital media, 5, 61
Digital philosophy, 11
Digital rights, 107
Digital Rights Managements (DRM), 9, 61
Digitalisation, 54
Discrete, 15

e-democracy, 108
e-government, 57, 108
e-voting, 107, 109, 111
Embedded, 2
Episteme, 15
Equipment, 127
Ethics, 24
Eurofighter typhoon, 3

Everyday computational, 14
Eyjafjallajökull volcano, 7

F16 Fighting Falcon, 3
Facebook, 1, 6, 109, 120, 165
Familiarity, 124
Financialisation, 155, 156, 161
Flow-chart, 113
Fly-by-wire, 3
Folksonomies, 60
Fordism, 36
Free software, 5, 62, 111
Function, 56

Geo, 6, 151, 163, 165
Gelassenheit, 140
German Idealism, 19
Glass-box testing, 67
Google, 1, 6, 55, 111, 118, 122, 125, 135, 169
GPS, 6, 122, 123, 165
Graphic user interface (GUI), 4, 5, 37, 136

Hacking, 32, 35, 65, 70, 89, 137
Hammer, 130, 135
Hardware, 104
Heidegger, Martin, 2, 13, 27, 123, 126, 141, 154, 161, 167
Hello, world!, 64, 84, 95
Hermeneutic(s), 86, 107
Heteronomy, 9
High frequency trading, 159
HTML, 18, 118
HTTP, 59
Humanities, 24, 118
Hybridity, 24
Hypertext, 139

IBM, 47, 51
Imaginary, 143
Immaterial labour, 40
Information, 59, 122, 143
Information literacy, 20
Information society, 60, 62
Innovation, 49
Instrumental rationality, 13
Intellect, 20
Intellectual property rights, 5, 39

Intelligence, 20
Interactivity, 115, 142
Interface, 97, 99
Internet, 4, 58, 64, 74, 117, 134, 142, 165
iPhone, 53, 79, 132, 149, 158, 162
iPad, 132
iPod, 15, 50, 53, 79, 122, 132

Jailbreak, 53, 79, 132
Journalism, 27

Kant, Immanuel, 18, 27
Kuhn, Thomas, 21, 131

Labour, 39, 61
Lifestream(s), 147, 155, 162, 166
Linux, 64
Literacy, 5, 27
Literature, 19
Location, 6
Logic, 47
Long-tail, 60
Lovelace, Ada, 47

Machines, 1
Manifest image, 129, 131, 136, 140, 168
Marx, Karl, 124
Materiality, 32, 62, 65, 75, 77, 85, 100, 151, 154, 166
Media
  Archaeology, 4
  Genealogy, 4
  Mechanology, 4
  Studies, 5
Mediation, 9, 16, 38, 109, 117, 127, 166, 168
Medium, 10, 35, 155, 163, 167
Medium theory, 5, 152
Memory, 152, 154, 171
Metaphor, 46
Methods, 56
Microblogging, 163
Microcode, 45
Microsoft, 67, 68
Mill, John Stuart, 131
Miwa, Masahiro, 95, 99
Mobile, 48

Mobile phone, 120, 123, 132
Moral depreciation, 42, 112
Multitasking, 134, 149
Music, 102

Napster, 59
Narrative, 26, 100
Network, 58, 62, 97, 99, 170
Network neutrality, 61
Newman, John Henry, 19
Nietzsche, Freidrich, 151

Obfuscated code, 75, 82
Object-oriented philosophy, 132
Object oriented design, 34
Object oriented programming, 56
Ontology, 43, 129
Ontotheology, 27, 129, 131
Open source, 5, 45, 61, 73, 86, 111
Oracles, 12

Patterns, 9, 26, 170
Parasite, 170
Peer-to-peer, 59, 61
Phenomenology, 119, 121
Philosophy, 18, 28
Plasma, 152
Platform, 56, 119
Platform studies, 4, 97
Plato, 120
Poetry, 30, 49
Political economy, 6, 39, 62, 113
Politics of code, 8, 107
Post, Emil, 47
Posthuman, 158
Post-Fordist, 36
Present-at-hand, 127, 129, 130, 134, 135
Privacy, 7
Processor, 38, 46, 104
Profit, 150
Program(ming), 46, 101
Protocol, 62, 165
Pseudocode, 52
Public, 144
Public sphere, 27, 108

Raymond, Eric, 73
Reading code, 9

Ready-to-hand, 127, 129, 130, 134, 155
Real-time, 7, 60, 164
Reason, 18
Redact, 76
Regime of Computation, 11, 139
Regulation, 9
Release candidate, 67
Reverse-simulation music, 100
Riparian, 144
Risk, 160
Rushdie, Salman, 144

Schmidt, Eric, 8, 55, 136
Scientific image, 129, 140
Screen essentialism, 36, 65, 137
Screenic, 48, 167
Search, 7
Search engine, 6
Search neutrality, 8, 9
Sellars, Wilfred, 121
Shannon, Claude, 47
Skill, 132
Social media, 6, 18
Social science(s), 21, 23, 118
Society of code, 66
Socio-technical device, 15, 62, 135
Sociology, 5, 24, 66
Socrates, 120
Software, 2, 4, 5, 15, 25, 31, 32, 39, 53, 64, 68, 97, 136, 137, 161, 167, 169
  Breakdown, 40
  Error, 41, 44, 69, 80
  Logics, 42
Software engineering, 32, 65, 114
Softwareized, 18
Software avidities, 6
Software engines, 4
Software studies, 4, 113
Software work, 32
Source code, 29, 52, 64, 73, 75, 81, 86, 92, 113
  Windows 2000, 68, 72
Speculative philosophy, 131
Speculative realism, 132
Squarciafico, Hieronimo, 120
Standing reserve, 2, 171
Stream, 14, 37, 53, 104, 134, 142, 143, 150, 152, 159, 160, 165, 170, 171
  Computational, 145

Data, 144, 155, 170
  Real-time, 108, 142, 164, 166
Structure of feeling, 6
Subscopic, code, 33
Super-medium, 10

Tagging, 60
Tarde, Gabriel, 24, 66, 164, 166
Techne, 15, 108
Technical device(s), 41, 63, 108, 125, 130, 149, 158, 169
Technical system, 120
Technological determinism, 119
Television, 135, 167
Temporality, 5, 97
Tests, 66
Test case, 65
Trial of strength, 65, 67, 82
Turing, Alan, 14, 20, 47, 125
Twitter, 109, 151, 155, 162, 169

UML, 66
Underhanded C contest, 75
Universitatis, 21
University, 18, 26
  Of Excellence, 19
  Post-modern, 19
  Of East Anglia, 72
Unreadiness-to-hand, 133, 141, 167, 168

Viacom, 8
Vicarious, 132, 137
Virtual, 4
Voter, 113

Wall-mart, 1
Waterfall model, 66
Web, 4, 59, 122
Web 2.0, 56, 67, 142
Web science, 5
Williams, Raymond, 119
Writing code, 9, 15

XOR gate, 102

YouView, 50

Zuse, Konrad, 47

Made in the USA
Middletown, DE
18 July 2023

35391692R00126